Fascination Astronomy

Arnold Hanslmeier

Fascination Astronomy

A cutting-edge introduction for all those
interested in the natural sciences

 Springer

Arnold Hanslmeier
Universität Graz
Graz, Austria

ISBN 978-3-662-66022-5   ISBN 978-3-662-66020-1   (eBook)
https://doi.org/10.1007/978-3-662-66020-1

This Springer Spektrum imprint is published by the registered company Springer-Verlag GmbH, DE, part of Springer Nature.
The registered company address is: Heidelberger Platz 3, 14197 Berlin, Germany

Preface

Astronomy is a science that pushes the boundaries. One has to deal with unimaginably huge dimensions of space and time, unimaginable hot and also cold objects, and even in our solar system, which is now also well explored by satellite missions, there are constantly new discoveries. In addition, astronomers cannot work directly with their objects of research. The only information we get about stars and galaxies is their radiation and position in the sky. Nevertheless, the laws of physics allow us to get information about these objects: Of many stars and galaxies we know what they are made of, how old they are, how far they are from us, etc. Findings from astronomy have had a significant impact on our thinking. The earth is by no means the center, but only a tiny planet in the universe, which is infinite, yet finite, and which itself has no center. In this book, which originated from a lecture given at the University of Graz for students of all faculties, we will convey astronomical knowledge without going into too much physics and mathematics. Somewhat deeper formulas and passages are separated from the rest of the text and can be skipped without losing the context. Nevertheless, the impression should be given that astronomical numbers are provable and the result of careful measurements.

In the first chapter we describe the fascination of the origin of the universe. Modern insights of theoretical physics help here and astrophysics and physics together bring new insights and raise new questions at the same time. We cover dark matter, the solar system, the sun, and the evolution of stars. Huge supermassive black holes are located in the centers of galaxies, and for the first time it is possible to detect them by observation. For more than 25 years we also know planets outside our solar system. Thus, at the end of the book we try to give an answer to the probably most exciting question of the natural sciences: Are we alone in the universe?

The book is intended not only for students but also for interested laymen, as well as for anyone who is interested in modern findings in natural science. Physics, especially astrophysics, can be extremely exciting, I hope my readers gain this impression by reading this book!

This third edition of this book takes into account the latest findings from satellite missions, such as the first successful soft landing on a comet or spectacular images of the dwarf planet Pluto, and recent missions to the Moon and Mars. Particularly exciting was the first direct observation of gravitational waves, which Einstein predicted but did not himself believe would ever be detected. Recent findings in exoplanet research and examples of exoplanet systems are discussed. Errors have been corrected and numerous new illustrations introduced to aid understanding.

Graz, Austria   Arnold Hanslmeier

Acknowledgments

I would like to thank Ms. Bettina Saglio and Ms. Meike Barth from Spektrum Verlag for their excellent cooperation. I thank my partner Anita for numerous discussions and her understanding and patience.

Graz, December 2022.

Contents

List of Tables

The Forces that Shape the Universe

1

1.1 Gravity

1.1.1 Newton and the Apple

Sir Isaac Newton (1643–1727, Fig. 1.1) is considered one of the most important physicists. In addition to pure physics, he was also concerned with optics, mathematics and other disciplines. His main work is the *Philosophiae Naturalis Principia Mathematica,* often referred to as *Principia for* short (Fig. 1.2). The work was published in 1687 in Latin, the language of science commonly used at the time. In this work, Newton presented the law of gravitation. He recognized that an attractive force acts between two masses. But how can this force be described, on what does it depend? The reasoning is very simple:

- The larger the mass, the stronger the force it exerts on other masses.
- The further the distance between two masses, the weaker the force. Experiments showed that the force between two masses decreases with the square of their distance.

Let us consider a few examples. All masses in the universe attract each other. An apple lifted into the air attracts the earth, and the earth attracts the apple. However, since the mass of the Earth is much greater than that of the apple, the apple falls to the Earth and not vice versa. Let us consider the earth-moon system. The Moon has only about 1/81 of the Earth's mass; therefore, the Moon can be said to orbit the Earth. In reality, both celestial bodies move around their common center of gravity, which, however, is inside the Earth because of the Earth's greater mass. The mass of the Sun is 333,000 times the mass of the Earth. Thus the earth orbits around the sun.

Fig. 1.1 Sir Isaac Newton.
(© Fine Art Images/Heritage
Images/picture alliance)

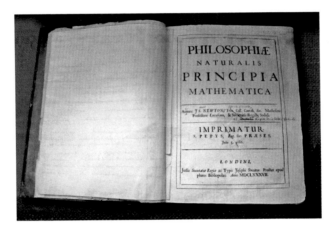

Fig. 1.2 Newton's main work in which he introduced the law of gravity. (A. Dunn, Creative
Commons License)

Digression One can write Newton's law of gravity as follows:

$$F = G\frac{M_1M_2}{R^2} \tag{1.1}$$

Here $G = 6.67 \times 10^{-11}$ is the gravitational constant, M_1 stands for the first mass, M_2 for the
second mass and R for the distance between the two masses. If one wants to calculate the
force (in the physical unit Newton, N), then the mass in kg and the distance R in m must
be used.

It is said that Newton came up with this law while he was under an apple tree and
thought that it was the same force that makes an apple fall to earth as the force that makes

the moon revolve around the earth. What seems obvious to us today was by no means so easy to accept in Newton's time. Newton's assumption presupposes that the same laws of nature apply on earth and in the *heavens*, as they were called at the time. Incidentally, the story of how Newton arrived at his law is also typical of the custom in today's scientific scene. You have to pick up ideas at the right time and then sell them largely as your own. Two years before Newton's publication, the astronomer E. Halley (considered the discoverer of the recurring comet every 76 years) sent Newton a treatise on the motion of bodies in orbit. And as early as 1674, the physicist Hooke suspected that there must be a force that binds the planets to the sun.

In this sense, Newton is not really the discoverer of the law named after him.

1.1.2   Where Does Gravity Stop?

So we can formulate somewhat abstractly the following: Gravity is a property that all masses possess. Gravity always has an attractive effect and what is interesting is that it cannot be avoided. Every mass exerts this force, it cannot be shielded. Of course, it would be very practical to shield gravity. Then we could float above the Earth's surface without propulsion. But what about astronauts who orbit weightlessly around the Earth in their space station? Does gravity no longer have an effect on them? It is true that the Earth's gravitational pull decreases with the square of the distance, but at an altitude of about 250 km, which is typical for spacecraft orbiting the Earth, this decrease is practically irrelevant. Why then are the astronauts and everything in the spaceship in what is always called a weightless state?

The answer is quite simple: the spaceship falls practically continuously around the earth. This is an accelerated motion, because the direction of the velocity changes continuously due to the circular orbit around the Earth. As soon as this acceleration is equal to the acceleration due to gravity acting at the height of the spaceship (i.e. gravity), one has the feeling of weightlessness.

So in principle we can change gravity by acceleration. When an airplane takes off, the passengers are pressed into the seats; when an elevator initially accelerates downward, we are somewhat lighter; when we free-fall, we also experience the feeling of weightlessness (although here we are slowed down by air resistance, so that after jumping out of the airplane, from a certain height, we fly towards the surface of the earth at a constant speed of 200 km/h).

1.1.3   How the Solar System Holds Together

We will describe the solar system in more detail in the following chapters. Here we are only concerned with gravity. The sun has about 99.8% of the mass of the whole solar system. So if all known eight planets (including earth), moons of planets, small planets (some 100,000 are known) and comets are added, their masses won't be more than 0.2% of sun. Therefore all these bodies have to move around the sun – exactly around the centre of gravity of the system, which however is variable due to different positions of the planets, but is always very close to the sun. By this motion, bodies experience centrifugal forces directed outward. This force is known and feared by everyone who has driven a car too fast into a curve. Our solar system is about 4.6 billion years old. The following equilibrium has functioned for almost all celestial bodies in the solar system for just as long:

Stable planetary orbit: centrifugal force = Attraction by sun.
That is why our solar system is relatively stable. The centrifugal force acts more strongly
the faster the body moves, and its amount increases as the radius of the orbit decreases.
If you drive into a very tight curve, you notice this.

Digression The centrifugal force is calculated from

$$F_{\text{centrifugal force}} = \frac{mv^2}{r},\tag{1.2}$$

where r is the orbital radius, m *is* the mass, and v is the velocity.

Let's look at the planet closest to the sun: Mercury. Because of its small distance from the sun, it experiences a much stronger attraction from it. Its distance is about 0.4 times the distance between the Earth and the Sun, i.e. the force with which Mercury is attracted by the Sun is equal to $1/0.4^2$ times the force with which the Earth is attracted. In order for Mercury not to crash into the Sun, it must move very quickly around it. The closer planets are to the Sun, the shorter their orbital period around it.

Around 1990 the first planets outside our solar system were found. Many of these planets move in very small orbits around their central star, often in only a few days.

1.1.4 From the Planetary System to the Universe

The universe has a hierarchical structure. Objects of the same kind always form structures and shapes that belong together. Planets do not occur in isolation, but in planetary systems. Stars arrange themselves into huge galaxies, often containing several hundred billion solar masses. Galaxies arrange themselves into galaxy clusters and the galaxy clusters in turn form superclusters. All these formations are held together by gravity. To prevent the stars of a galaxy from crashing into its supermassive black hole located at the center, they orbit around the center. Our Sun is about 30,000 light years away from the center of our home galaxy, the Milky Way, and takes more than 220 million years to orbit.

Due to gravitational forces, our cosmos is dynamic and by no means static. Also the fixed stars in the sky are moving. However, they are very far away and therefore the annual

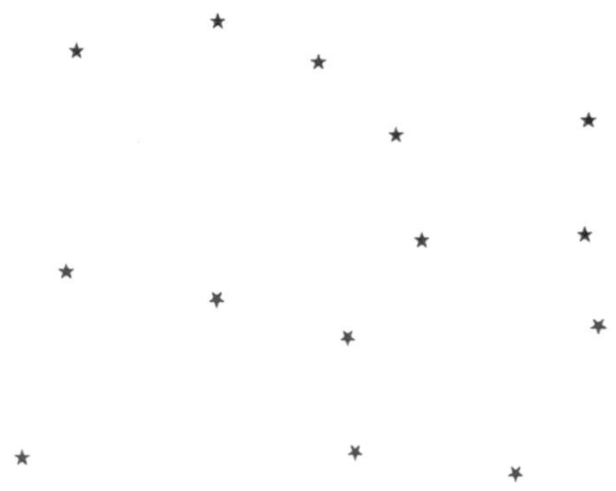

Fig. 1.3 The constellations' appearance slowly changes due to their proper motion. The Big Dipper as it appears today (red stars) and as it will appear in 100,000 years (blue stars). (A. Hanslmeier)

movement of the fixed stars in the sky cannot be measured with the naked eye. Only after some 10,000 years the positions of the stars in the sky shift noticeably even for the naked eye. So our ancient cultures saw practically the same constellations as we do. This is illustrated in Fig. 1.3 using the well-known asterism of the Big Dipper.

What Is Gravity? So gravity is omnipresent within universe, each mass affects such attracting force onto other masses. Because everything is in motion, the universe does not collapse. Where does this motion actually come from? And what is gravity really? Two answers to these questions, which are not really satisfying, but hopefully will become clearer as you read the book:

- A body with a mass M generates a gravitational field around itself. This statement can be found in classical physics books, but actually it doesn't get us anywhere.
- Gravity is a fundamental property of *space-time* and therefore cannot be switched off. Masses cause a local curvature of space-time. This is the statement of Einstein's general theory of relativity.

Apart from matter as we know it, is there anything else on which gravity acts? The answer is clearly yes. We cannot directly see, measure or observe this matter, but it acts through gravity. It is therefore called *dark matter*. We will come back to this later.

For the moment, let's just note that gravity determines the major structures of the universe, but there are still many unanswered questions.

1.2   The Electromagnetic Force

Gravity is not the only force in the universe.

Electromagnetic forces were already known to the ancient Greeks. When amber (Greek *electron,* hence the name electricity) is rubbed, charges are created that attract other smaller pieces of paper.

Gravitational forces determine the motion in the universe, electrical forces determine the motion of electrons around the atomic nucleus, the cohesion of molecules. However, they also play a role in astronomical objects. In the corona of the sun, which is several million degrees hot, magnetic field lines determine the motion of the plasma.

1.2.1   Charges

Besides the property of *mass*, there is another property of many bodies and particles, the *charge*. Atoms consist of a positively charged atomic nucleus and negatively charged electrons. Normally, the positive and negative charges are equal, and the atom as a whole is electrically neutral. However, in many chemical elements, the outer electrons are only lightly bound and can be displaced. If there are then too few negatively charged electrons in an atom, it is electrically positively charged; if there are more electrons than the positively charged protons in the nucleus, the atom is negatively charged. You can release electrons from an atom by adding energy. This is called ionization (Fig. 1.4).

Ionized atoms are therefore always found at high temperatures. As an example for this we consider our sun. At the surface, the ionization is not very high, because the temperature is only about 6,000 K. In the several million degrees hot corona of the sun, certain elements like iron can have lost up to 14 electrons. Astrophysicists then write FeXV for this.

In the classical picture, an atom consists of a nucleus with positive charges (protons) and neutral particles (neutrons) as well as the electrons in the shell (negatively charged). Electrically charged atoms are called ions.

In nature, charge does not occur at random, but always in multiples of the electric elementary charge e.

Charge occurs as a multiple of the elementary charge e:

$$1e = 1.6 \times 10^{-19}\,\mathrm{C}$$

Digression   An electron carries the charge $-e$, a proton the charge $+e$. The charge of a body is then always $Q = Ne$, N is the number of its charged particles. One measures the charge in the unit C, coulomb. An electric current is generated when charge Q is transported during a time t:

$$I = Q/t. \tag{1.3}$$

The current is given in amperes $= 1\mathrm{A}$. Let us assume that the charge is $Q = It = 1\mathrm{A}\,\mathrm{h}$. It is a multiple of the elementary charge, i.e. $Q = Ne$, and therefore $N = 3600/1.6 \times 10^{-19} = 2.25 \times 10^{22}$ electrons give the charge $1\mathrm{A}\,\mathrm{h}$.

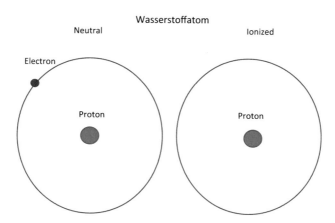

Fig. 1.4 The simplest atom is the hydrogen atom. It consists of a proton (positively charged) in the nucleus and an electron (negatively charged) in the shell. *On the left* is the neutral hydrogen atom (proton + electron), on *the right the* ionized hydrogen atom (only nucleus)

1.2.2   The Coulomb Force

Physics is simple, as is astrophysics, of course. Once you understand a principle, everything else is analogous. Coulomb's law is analogous to Newton's law of gravitation:

Property	Law of gravity	Coulomb Force
Applies to	Masses, m_1, m_2	Applies to charges q_1, q_2
Force decreases with	Distance squared	Distance squared
Reaches until	Infinite	Infinite
Is proportional to	Product of the masses	Product of the charges
Differences	Only one kind of masses	Positive and negative charges

Digression   The force acting between two charges q_1, q_2 is given by:
   Coulomb Force.

$$F = \frac{1}{4\pi\epsilon_0} \frac{q_1 q_2}{r^2}$$
(1.4)

$\epsilon_0 = 8.854 \times 10^{-12} \frac{As}{Vm}$ is the electric field constant.

However, we must reemphasize one important difference from the law of gravitation: The electric force between two charged particles can be:

- attractive: if both charges have different signs; e.g. electron and proton in the hydrogen atom.
- repulsive: if both charges have the same sign.

Otherwise everything as usual: Force depends on charges and decreases with square of distance.

1.2.3   Atoms: Miniature Planetary Systems

Let us consider the simplest and at the same time most common atom in the universe: hydrogen atom (Fig. 1.4). It consists of an atomic nucleus with a positively charged proton and a negatively charged electron orbiting it. Both would attract each other, but the circular motion creates an outward centrifugal force, analogous to the planets.

A peculiarity of nature, however, is that electrons can only orbit around the atomic nucleus at certain distances (one speaks of energy levels). This is only understood with the help of quantum mechanics and it contradicts our everyday experience. If we apply this to our planetary system, then planets could only occur at certain distances from the sun. Theoretically, a planet could be at any distance from the sun.

If electrons jump from one energy level E_1 to another E_2 and $E_2 > E_1$, then:

- Radiation is emitted when the electron jumps from the E_2 level to the E_1 level;
- Radiation is absorbed when the electron is lifted by incident energy from the lower level E_1 to the higher E_2

Thus, one can understand the formation of spectral lines. In case of the hydrogen line $H - \alpha$ an electron jumps from the third to the second shell (emission) or from the second to the third shell (absorption). The line is at a wavelength of 656.3 nm (1 nm $= 10^{-9}$ m) and can be seen in the red region of the spectrum (see also Fig. 1.5). Many gas nebulae therefore glow red, including the higher atmosphere of the Sun, as this line is formed there. An example is the gas nebula with the designation M20, the so-called Trifid Nebula. You can see the hydrogen gas glowing red (Fig. 1.6). This gas nebula, where new stars are formed, is about 5,200 light years away from us.

1.2.4   Electricity + Magnetism = Electromagnetism

We all know magnets, e.g. a horseshoe magnet (Fig. 1.7).

Magnets always occur bipolar, i.e. with a north and a south pole. So there are no magnetic monopoles (however these could be a remnant of an early phase of universe). Already at nineteenth century, especially Maxwell (1831–1879) did show, electric and magnetic appearances can be combined:

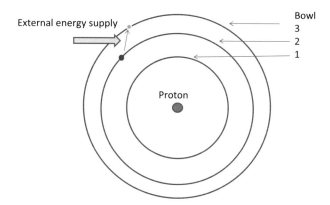

Fig. 1.5 Formation of the absorption line H − α by transition of the electron from state 2 to state 3 when energy is supplied from the outside

Fig. 1.6 Trifid Nebula M20, a nice example of red glowing hydrogen gas *(right)*. (A. Hanslmeier, Pretal)

Fig. 1.7 A horseshoe magnet attracting iron filings. Magnets are bipolar, so they have a north and south pole. (Wikimedia Commons; cc-by-sa 3.0)

- Changing electric fields generate magnetic fields,
- changing magnetic fields generate electric fields,
- Magnetic fields are created by currents (e.g. around a current-carrying conductor).

The magnetic field of a bar magnet is sketched in Fig. 1.8. You can see the two poles N and S.

The earth's magnetic field protects us from energetic particles from the cosmos (most of which come from the sun). Charged particles cannot penetrate magnetic fields and are deflected. Where does the magnetic field on Earth come from? The Earth's core is liquid and very hot. There are currents of charged particles there, which are known to produce a magnetic field. By a so-called dynamo process one explains how the earth's magnetic field renews itself again and again after phases of polarity reversals.

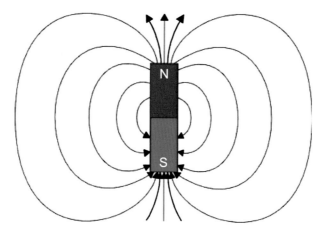

Fig. 1.8 Magnetic field of a bar magnet

1.2.5   Plasma: The Most Common State of Matter in the Universe

We all know the three states of matter: solid, liquid, gaseous. But there is also a fourth state of matter. The term plasma refers to a gas that is at least partially ionized. So there are high temperatures. Because of the ionization the gas is conductive and depending on the density of the plasma the movement of the plasma determines the structure of the magnetic field (this is the case near the surface of the sun) or the magnetic field determines the movement of a thin plasma. In Fig. 1.9 one can see these structures. In the corona of the Sun, which is several million degrees hot, the plasma structure is determined by the magnetic field lines. The corona has an extremely low density. These loops or arcs are also called magnetic loops.

About 99% of the visible matter in the universe is plasma. On Earth, we observe plasma states, for example, during electrical discharges.

1.3   The Strong and the Weak Force

1.3.1   What Holds Atomic Nuclei Together?

With our previous ideas of the structure of matter we get a problem: As soon as there is more than one positively charged proton in the atomic nucleus (all elements except hydrogen), the atomic nucleus cannot be stable. After all, the Coulomb repulsion forces act between the positively charged protons. Let's consider helium: It consists of two protons and two neutrons. The neutrons are electrically neutral, but the protons repel each other. Therefore, there must be a force that is stronger than the repulsion between the two positively charged protons. Although the neutrons are electrically neutral, they

Fig. 1.9 The arc-shaped structures in the Sun's corona are formed by the arrangement of the plasma along the magnetic field lines. Image: TRACE, NASA

hardly mitigate the repulsion. This force, which holds atomic nuclei together, is called the *strong force.*

The strong force is such that its strength increases with increasing distance, but then rapidly approaches zero. So if you move two protons far enough away from each other, they no longer feel anything from the strong force. Protons and neutrons are not elementary particles in the strict sense. They consist of smaller particles called quarks. The quarks are held together by gluons (see Fig. 1.10). Quarks have third-number multiples of the elementary charge *e.*

Let us consider a proton: This consists of three quarks, two so-called *u* (stands for *up*) and one *d* (stands for *down*) quark. The *u*− quark has the charge +2/3, the *d*− quark the charge −1/3. Thus the charge of the proton is $2/3 + 2/3 - 1/3 = 1$. The neutron consists of one *u*− quark and two *d*− quarks, so its charge is $+2/3 - 1/3 - 1/3 = 0$.

Quarks never occur as free particles, one speaks of a *confinement*. Only at the beginning of the universe, when temperature and density were extremely high, there was a quark-gluon plasma.

1.3.2 The Weak Force

This force is needed to explain the radioactive decay of the elements. Even free neutrons decay with a half-life of about 10 min. Figure 1.11 shows the *beta decay of* a neutron. The neutron *n* consists of three quarks, *u, d, d,* it decays via a W^- -particle into an electron e^- and a neutrino $\bar{\nu}_e$ as well as into a proton *p*, which consists of the three quarks *u, d, u. The neutron n decays with a half-life of about 11 min.*

The weak force plays an important role in inverse beta decay, which occurs, for example, during nuclear fusion in the interior of stars. In this process, hydrogen is

Fig. 1.10 Protons (sketched here) and neutrons each consist of three quarks held together by gluons

Fig. 1.11 The beta decay of a free neutron, essentially producing an electron and a proton. The W^- particle indicates the weak force

Fig. 1.12 The inverse beta decay

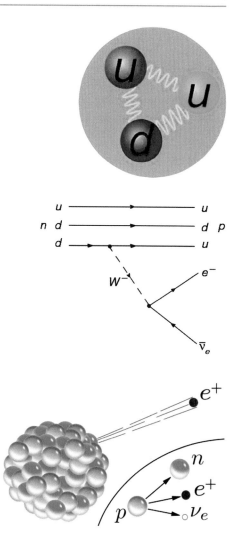

converted to helium in most stars. This happens in several steps, first two protons react with each other and form a deuterium nucleus. This contains a proton and a neutron created by the conversion of a proton. In this process, a *u quark of* the proton is converted into a *d quark*; here the weak interaction occurs, a W^+ particle is formed, and from this a positron and a neutrino are formed (Fig. 1.12). In both cases, the W particle is found, which is characteristic of the weak force.

1.4   Elementary Particles

We have described forces that act in the universe. In physics, instead of these four basic forces, we also speak of interactions.

Matter consists of atoms. These in turn are made up of an atomic nucleus and a shell. In the atomic nucleus we find the positively charged protons p as well as the neutral neutrons n. These two particles are also called nucleons.

Examples
- Hydrogen atom: The neutral hydrogen atom consists of a proton in the nucleus and an electron in the shell. However, there are several isotopes, which are atoms with the same number of protons but different numbers of neutrons. The hydrogen isotope deuterium contains one proton and one neutron in the nucleus, the hydrogen isotope tritium contains one proton and two neutrons in the nucleus.
- Helium: Helium atoms always contain two protons. The most common helium isotope contains two protons and two neutrons in its nucleus.

For a long time it was believed that these nucleons are real elementary particles, i.e. they cannot be further decomposed into smaller particles. Today we know that this is not true. Protons and neutrons are made of the already mentioned quarks. These cannot be further split, and hence the name elementary particles. Another class of elementary particles are the leptons, which include the electrons found in the atomic shell.

Quarks and leptons are therefore elementary particles of which all matter known to us is composed.

1.4.1   Interactions

The universe would be completely uninteresting if these elementary particles existed without interacting with each other. So there are the four forces or interactions between the elementary particles already discussed. In the sense of modern physics, one speaks of *field quanta,* which transmit forces. For each interaction there are own field quanta.

1. Gravitation: acts between masses; the field quanta that transmit them are called *gravitons.* The range of gravitation is to infinity. The strength is 10^{-39} (one refers to the strength relative to the strong force). However, the graviton has not been experimentally proven to date. Another prediction of general relativity are gravitational waves, disturbances of space-time propagating at the speed of light. Such waves are emitted by accelerating masses. In early 2016, such waves, emitted when two massive black holes merged, were detected for the first time. The two black holes were more than a billion light-years away from us and contained 29 and 36 solar masses, respectively. This detection opens a new window for modern astrophysics.

2. Electromagnetic interaction: Field quanta which transmit this interaction are called *photons*. This interaction also extends to infinity. One can easily convince oneself by the following simple experiment that this interaction must be much stronger than gravity. Imagine how a piece of iron is pulled upwards by a magnet in spite of gravity. The strength of the electromagnetic interaction is 10^{-2}. But since there are two kinds of charges, the universe is electrically neutral in the grand scheme of things.

3. Weak interaction: The field quanta are called W^+, W^-, Z^0 particles; this force has a very short range of only 10^{-18} m. The relative strength is 10^{-14}.

4. Strong interaction: *Gluons* are the carriers of this interaction. It is only effective in the region of atomic nuclei. The range is 10^{-15} m. For quarks, the range is assumed to be infinite. Quarks can therefore not be observed as free particles, at least in today's universe.

In the sense of this theory, forces are transferred between particles through the exchange of field quanta. As an analogy, imagine two ice skaters throwing snowballs at each other. Field quanta are constantly absorbed and emitted.

In the subatomic realm, the strong and weak interactions dominate, electromagnetic forces determine the structure of atoms and molecules, but the structure of the universe is determined by the weakest force, gravity.

1.4.2 Elementary Particles: Description

Nowadays the following model is used:
There are three groups of elementary particles:

1. Leptons,
2. Quarks,
3. Field quanta.

For every particle there is also an antiparticle.

Leptons The Greek word *leptos* means light. We are talking about light particles. Electron and its antiparticle, the positron, are certainly known to all. Other leptons are the muons and tauons, but they decay into electrons and neutrinos.

One can represent such processes in a *Feynman diagram*, as *y-axis* one uses the time T, as *x-axis* the location of a particle. Antiparticles are represented as particles that run inverted in time. As an example, Fig. 1.13 shows the decay of a μ^--myon. It decays via the weak interaction into a W^- particle as well as an antielectron neutrino $\bar{\nu}_e$ and an electron e^- as well as a ν_μ, a muon neutrino. The lifetime of a muon is less than one millionth of a second.

Fig. 1.13 Decay of a muon

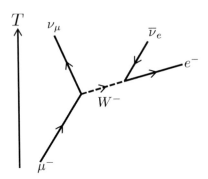

Muons are created when cosmic ray particles arrive in the Earth's high atmosphere. The relativistic effect of time dilation is noticeable in the case of extremely fast-moving muons. Due to their short lifetime, half of the muons produced in the atmosphere would already have decayed after a distance of 450 m, so they would never reach the Earth's surface. Fast muons, however, decay more slowly, which is why they can also be observed near the Earth's surface.

Example
Muons travel at 99.5% of the speed of light. How large does their lifetime appear from an observer resting on the surface of the earth? The time dilation is determined from the formula:

$$T' = \frac{T_0}{\sqrt{1 - \frac{v^2}{c^2}}} \qquad\qquad (1.5)$$

with $T_0 = 10^{-6}$ s follows: $T' = 10 \times T_0$ so the observer measures about 10 times the lifetime. We see: So travel keeps young.

Neutrinos are uncharged particles with a very low rest mass. They only react to the weak interaction. There are probably about 10^9 times more neutrinos in the universe than nuclear particles. Therefore, despite their very small mass, they are essential for the total mass in the universe.

Quarks Where the word quark comes from is no longer known exactly, but possibly from the book *Finnegans Wake* by J. Joyce, where the sentence is dropped: "Three quarks for a master mark" (three cheese highs make a whole man). These are heavy particles that never occur singly. We know of 18 different quarks (up, down, strange, charmed, bottom and top, abbreviated: u, d, s, c, b, t). Each of these quarks occurs in three versions, which are called red, green and blue. They are also referred to as color charges. Of course, these are purely fanciful names and only describe quantum numbers. For each quark there exists also an antiparticle.

Field Quanta So field quanta are the transmitters of forces. The graviton could not be proved up to now.

1.4.3 Quarks and Hadrons

So how do the quarks form the hadrons? The Greek word *hadros* means strong, for hadrons the strong interaction is essential. One distinguishes with the hadrons again between

1. baryons: Examples are protons and néutrons. They consist of three quarks: the proton consists of (uud), the charge of a u quark is +2/3e, that of a d quark $-1/3e$. Therefore, the charge of the proton is *e*. Neutrons consist of (udd), which gives the charge zero. Other examples of hadrons include the sigma particles $\Sigma^+, \Sigma^0, \Sigma^-$ which consist of (uus), (uds) and (dss) respectively.
2. Mesons: These consist of a quark and an antiquark. As an example we consider the pions π^0, π^+, π^- , which consist of $(u\bar{u})$, $(u\bar{d})$, $(\bar{u}d)$. There are 30 other mesons known

1.4.4 We Build a Universe

So we have everything we need to create a universe. Particles as well as interactions, that is, forces acting between particles. We have discussed the *standard model of* physics used today, which describes everything. Admittedly, this model does not look particularly simple. There are a bewildering number of particles. The question arises whether it could not all be simplified. This is exactly the goal of modern physics, a theory of everything, called Theory of Everything, TOE. To understand how such a simplification might be imagined, we need to look at the early universe. In the extremely hot and dense conditions that existed tiny fractions of a second after the Big Bang, we can imagine that all forces were united into a super force, or particles were formed from the high energies. This is the content of the next chapter.

The Big Bang: How It All Began

2

In this chapter we describe the early phases of the universe. First, we consider what evidence there is from observations that the Big Bang really happened. Then we go back to the standard model of the physics of particles and forces described in the previous chapter.

You will be able to answer the following questions after reading this chapter:

- When and how did the universe come into being?
- What was before the universe?
- What is outside the universe?
- What does space-time mean?
- What is the core statement of general relativity?
- How do we know there was a big bang?

2.1 The Galaxy Escape

2.1.1 Measuring the Universe

Around 1900, the picture we had of our universe was relatively simple. The Earth, and thus the solar system, is part of a vast system of about 100 billion stars called the Galaxy or Milky Way. The first ideas about the true nature of the Milky Way were already held by the ancient Greeks. Democritus (about 460–370 BC) thought that the Milky Way was really a collection of distant stars that we perceive as nebulae. By 1900, larger telescopes of more than a meter in diameter were available and numerous nebulae were known. Many of these

A. Hanslmeier, *Fascination Astronomy*,
https://doi.org/10.1007/978-3-662-66020-1_2

Fig. 2.1 Determining the distance of a star by measuring its annual parallax sin$p = a/r$, a ... Earth's orbital semi-axis, r ... distance of the star

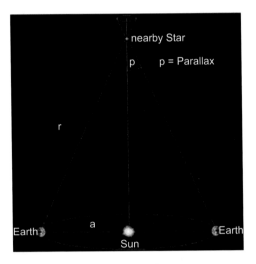

nebulae showed a spiral structure. A great dispute arose about the true nature of these nebulae:

- Are all nebulae real gas nebulae, similar for example to the well-known Orion Nebula?
- Are all nebulae so-called world islands, i.e. systems of stars similar to the Milky Way but independent of it?

This question can only be answered when the distance to these nebulae is known. Distances of relatively close stars could be determined already around 1850 by the method of the annual parallax (Fig. 2.1). As a result of the annual motion of the Earth around the Sun, the position of a nearer star shifts relative to distant background stars over the course of a year. This angle is very small, being less than one arcsecond in the sky (one arcsecond, 1/3600 of a degree). Let us imagine a 1 EUR coin at a distance of 2 km. Then you can see it at an angle of one arcsecond!

Only extremely precise measurements allowed the determination of the parallax of the nearest stars.

The first measurement of a star parallax was made by Friedrich Wilhelm Bessel in 1838. He chose the relatively fast moving star 61 Cygni. Its parallax is only 0.3 arc seconds. For comparison: The apparent radius of the moon in the sky is about 900 arc seconds.

But this method is not sufficient for large distances, because with increasing distance the angle becomes smaller and smaller. However, a class of variable stars was discovered, the Cepheids. These periodically change their brightness by pulsation, the stars inflate and contract again (Fig. 2.2). It was highly interesting when it was discovered that there is a relationship between the (easily measurable) period of the change in brightness and the true luminosity of the Cepheids. If one compares the true luminosity of a Cepheid determined from the period with the measured luminosity, its distance follows immediately. Why? Consider that the brightness measured by us on Earth naturally depends on the distance of the star.

Fig. 2.2 Periodic change in brightness of a Cepheid

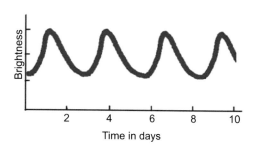

So to determine the distance of the nebulae, all you have to do is find a Cepheid in that system, determine the period of the change in brightness (it is only a few days), and you get the distance of the nebula in which that star is embedded.

2.1.2 Hubble

E. Hubble (1889–1953, Fig. 2.3) tried to determine the distance of the Andromeda Nebula with the largest telescope ever built by man on Mt. Wilson at that time. The telescope is constructed as a reflecting telescope, i.e. one has a mirror of about 2.5 m diameter as a collecting surface for light. With this powerful instrument Hubble was able to find Cepheids in the Andromeda Nebula (Fig. 2.4) for the first time and thus determine the distance of the nebula to us: The value measured by Hubble was 700,000 light-years. This value is wrong, the modern value is about 2.5 million light years, but nevertheless it became immediately clear: The Andromeda Nebula is an independent star system similar to our Milky Way. It was only at this point that people began to imagine how immense the expansion of the universe must be.

In astronomy, the light year is used to measure distances. What is a light year? Light travels at a speed of 300,000 km per second. This is slightly less than the distance between the Earth and the Moon. In one year, light travels about 10,000,000,000,000 km, or 10^{13} km.

Light

Spread:	300, 000 km/s
	In 1 s about eight times around the earth!
Earth-moon:	1 s
Earth-sun:	8 min
Earth-planets:	Several minutes to hours
Earth-nearest star (α Centauri):	4.3 years

Hubble showed: There are galaxies outside our system; he determined the distance of the Andromeda Galaxy for the first time.

Fig. 2.3 E. Hubble, the
discoverer of galaxy escape.
(Observatories of the Carnegie
Institution of Washington)

Fig. 2.4 The Andromeda Galaxy M31, the closest galaxy to us, is already visible with the naked eye
or binoculars under very good conditions. (Amateur photo)

2.1.3   A Look into the Past

Before we delve deeper into Hubble's most famous discovery, one more important note. Light travels at a finite speed. It takes about 8.3 min to travel from the Sun to the Earth. When we observe the Sun, at the moment of observation we see light that was emitted from the Sun 8.3 min ago, because that is the time it takes for light to travel to the Earth.

Light   Light propagates at 300.000 km/s. It takes 8.3 min to reach the earth, which is $8.3 \times 60 = 498$ s. Thus the distance earth-sun is 498 s $\times$ 300.000 km/s~150.000.000 km.

A telephone call to a colleague on Mars would be a laborious affair. It takes several minutes for our question to reach them, as radio signals also travel at the speed of light at most. And we have to wait at least twice as long for an answer. This becomes even more noticeable with distances between stars. The nearest star (apart from the Sun) is Alpha Centauri, about 4.3 light years away. So a radio signal there takes 4.3 years to travel, and if we send a message today, we will have to wait at least 8.6 years for a response, assuming there is someone there who understands us and responds. Moreover, we are now observing radiation from this star, which was emitted 4.3 years ago. Since the light from the Andromeda Galaxy has been traveling to us for more than 2.5 million years, we are observing radiation from this galaxy with our modern telescopes that was emitted when there were no humans on Earth! The radiation of an object that is, for example, 5 billion light years away comes from a time when there was no Earth and no Sun in the universe.

Because of the finite propagation speed of light (300.000 km/s), a look into the depths of the universe is also a look into the past.

2.1.4   Galaxy Escape

Back to Hubble. After measuring the distance of the Andromeda Galaxy, he analyzed other galaxies. At the same time, he also studied the spectra of these galaxies. In a spectrum, the radiation from an object is dispersed into its individual colors, and dark lines are seen, usually from certain chemical elements of that star. These dark lines are also called absorption lines. They are formed during electron transitions in the atom. To great surprise, it turned out that practically all galaxy spectra have lines shifted to red. A red shift of the lines can be interpreted by the *Doppler effect*. If a radiation source (star or galaxy) moves away from the observer, then its lines appear shifted to red. This is explained in Fig. 2.5 by means of a sound source approaching observer 1 and moving away from observer 2. There is an increase in frequency or a reduction in wavelength; this applies analogously to light waves.

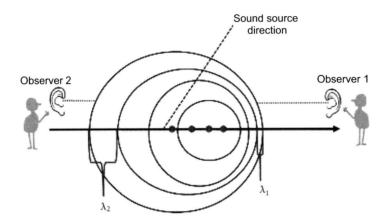

Fig. 2.5 Sketch explaining the Doppler effect. For a sound source moving towards observer 1, the audio frequency is increased, the wavelength λ_1 is decreased, from observer 2 the audio frequency is decreased, the wavelength λ_2 is increased. (Credit: University of Bayreuth)

Digression From the wavelength shift $\Delta\lambda$ one can immediately determine the velocity from the well-known Doppler formula:

$$\boxed{\frac{\Delta\lambda}{\lambda} = \frac{v}{c}}, \tag{2.1}$$

Where $c = 300.000$ km/s is the speed of light and λ is the wavelength of the unshifted line.

The Doppler effect can be heard with sound waves. If an emergency vehicle with its horn switched on approaches the observer, the sound appears to be higher; if it is further away, the sound becomes lower. From the redshift of the spectral lines of galaxy spectra (of course, this applies to all objects), Hubble was thus able to determine what velocities they possess. In addition to the measured redshift for all galaxies mentioned above, Hubble had the idea of plotting the velocities against the distance of the galaxies. This revealed a simple relationship: the speed at which a galaxy moves away from us depends on its distance, the further away, the faster. This is the famous Hubble law:

Hubble's Law:

$$\boxed{v = rH}. \tag{2.2}$$

H is the Hubble constant . This gives us a very simple way to measure the distance r of a galaxy. One simply determines the speed v with which it moves away from us.

The fact that all galaxies are moving away from us is called galaxy escape. Galaxy motion can be determined by the Doppler effect.

2.2   The Expansion of the Universe

2.2.1   Are We the Centre?

Again and again, people tried to explain the universe through a world model or world system. In the *geocentric* system, one imagines that the earth rests at the center of the cosmos, and that the sun, moon, planets, and other stars move around it. This is also the impression we get from the apparent daily motion of the stars. We still talk about the sun rising in the east and setting in the west. The geocentric system becomes complicated when we look more closely at the motion of the planets. Their orbits show strange loops in the sky at certain times, which are very difficult to understand by assuming a motion around the earth.

Figure 2.6 shows how the looping motion of a planet lying outside the Earth's orbit can be easily explained. We assume a *heliocentric* system, i.e. the Sun is at the centre and the Earth and the other planets orbit around it. As we saw in the previous section, the orbital velocity of a planet decreases with its distance from the Sun. Whenever the Earth overtakes the outlying planet in its orbit, we observe a looping motion of the planet in the sky.

However, if one assumes that all planets move around the resting earth, such a movement can only be explained with a complicated epicycle theory, according to which the movement of the planets can be represented in principle by superimposing movements on circles. The disadvantage of this explanation is, it becomes very complicated. As early as ancient Greece, Aristarchus (310–230 BC) first suggested that the Sun, rather than the Earth, might be at the center of the universe. This is called the heliocentric world system (Fig. 2.7).

Only with N. Copernicus this system became known in Europe – he published his main work *De revolutionibus orbium coelestium in* the year of his death 1543. With this the earth moved away from the center of the universe, a fact unacceptable for the doctrine of the church at that time. Famous from this time is the trial against Galilei (1564–1642). In 1624, Galileo traveled to Rome six times to discuss his positions regarding the heliocentric world system with Pope Urban VIII. The Pope encouraged Galileo to continue his research, but to emphasize that the heliocentric world model was only a mathematical hypothesis. So in 1632 *Galileo* published the *dialogo*, where the two world systems were discussed in dialogue form. The representative of the geocentric world system, Simplicio, does not come off very well in it, however. In 1633 Galileo was summoned to Rome, had to renounce his theses and was banished to lifelong house arrest. So Galileo was not burned at the stake like G. Bruno (in 1600), but he was grounded, even though the Pope of the time was one of his friends. One should not always rely on friends. The quotation attributed to Galileo *eppur si mouve – and it (the earth) moves nevertheless does* not come with very high probability from him. Incidentally, the verdict against Galileo was passed by ten cardinals, three of whom voted against guilty. It was not until 1992 that Galileo was officially rehabilitated by the Roman Catholic Church under Pope John Paul II.

Fig. 2.6 Sketch explaining the looping motion of an upper planet whose orbit lies outside Earth's orbit. (CC-BY-SA3)

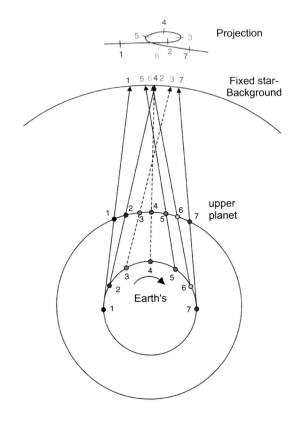

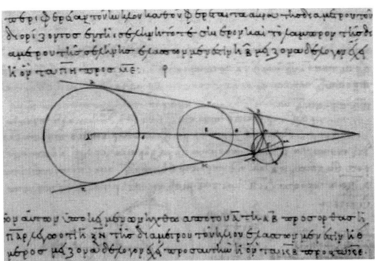

Fig. 2.7 Sketch from the traditions of Aristarchus. In it he shows how the ratio of the distance Earth-Moon to Earth-Sun can be determined from simple angular measurements. (Library of Congress)

But the development goes much further. The final proof of the correctness of the heliocentric world system was given when one could measure the annual fixed star parallaxes, that is, only 300 years after the appearance of the theory. The daily movement of the stars in the sky can be explained quite simply by the rotation of the earth. Stars do not rise in the east and set in the west, but the earth rotates from west to east during this time. Around 1900 it was known that the sun and thus the solar system is about 30,000 light years away from the center of the Milky Way, and about 20 years later Hubble discovered the escape motion of the galaxies. Have we thus once again moved to the centre of the universe?

2.2.2 The Universe Expands

The galaxy escape can again be explained quite simply if one assumes that all space is expanding. Imagine a balloon. We mark small points on it and blow up the balloon. No matter which point you then start from, you always have the impression that all other points are moving away from this point. This is exactly what is happening to the universe, it is expanding. Also, as an inhabitant of the Andromeda Galaxy, or any other random galaxy, one would have measured a Hubble law and therefore have the impression of galaxy escape. Thus the position of the solar system, and therefore that of the Earth, in the universe has again become relative; we are not at the centre. In Fig. 2.8 this effect is sketched again.

So where is the center of the universe? We go back to our model of the balloon being inflated. The universe resembles the ever increasing surface of the balloon. Where is a centre on this surface, where is an excellent special point on such a surface at all? The solution is simple: There are no excellent points, neither on the balloon nor in the universe, there is no center or special point in the universe.

Every point in the universe is equivalent. On large scales of a few hundred million light years, the universe appears to be homogeneous and isotropic. Isotropy means that the appearance of the universe on large scales is independent of the direction in which we look.

▶The correct interpretation of galaxy escape is therefore: The universe is expanding. It has no center.

2.2.3 The Age of the Universe

Let's go back to the example of the inflated balloon. If we inflate evenly, then it is easy to calculate back when we started to inflate.

Digression This can also be seen in Hubble's law by writing the units: Velocity in km/s, Distance in km:

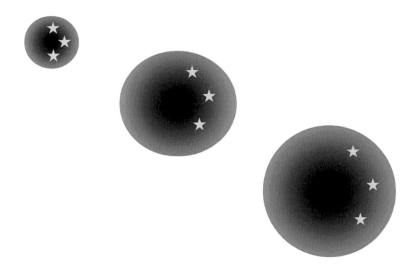

Fig. 2.8 Sketch of the expansion of the universe, i.e. of space-time

$$v = rH \quad [\text{km/s}] = [\text{km}]H, \tag{2.3}$$

from this the unit of length km is shortened and we get

$$1/s = H. \tag{2.4}$$

The reciprocal of the Hubble constant 1/H therefore has the physical dimension of a time, this is called the Hubble time. This is a measure of the age of the universe.

If we assume a uniform expansion, then we get a world age of about 13.7 billion years. This also agrees well with other observations.

The further we go back in the history of the universe, the smaller it becomes, because the expansion becomes smaller and smaller. 13.7 billion years ago, the universe was extremely small, extremely hot, extremely dense. It developed through an explosive process, for which the term *Big Bang* was introduced.

An important proof for the correctness of the big bang theory is therefore the observed expansion of the universe.

But there are two other important pieces of evidence for the correctness of the Big Bang theory.

2.3   The Hot Big Bang

2.3.1   The Universe and the Refrigerator

Let's remember the refrigerator principle. A refrigerator works by compressing a coolant strongly by a compressor and then expanding it. During this expansion, it cools down. So a gas that was previously highly compressed and then expanded cools down. The reverse is true when we inflate a bicycle inner tube. The compression of the air increases the temperature.

Digression   Let us assume that before inflation the temperature of the air in the hose was T_1, then after inflation the temperature becomes T_2. Before inflation the air volume is V_1, after inflation V_2, then an adiabatic change of state applies – as you can read in any physics textbook:

$$\frac{T_1}{T_2} = \left(\frac{V_2}{V_1}\right)^{\kappa - 1}.$$

(2.5)

In this formula, κ is the adiabatic exponent, which depends on the properties of the gas. The main point here: If the volume of a gas changes, then:

$$T \sim \frac{1}{V}.$$

(2.6)

It follows: The larger the volume, the lower the temperature. Let us now apply the refrigerator principle to the universe. In the beginning, the universe was very dense, so the volume was very small. Since temperature is inversely proportional to volume, the temperature in the early universe must have been very high. But the universe expanded, so the volume increased and the temperature decreased. Physics can be that simple!

The young universe was hot and slowly cooling.
So we expect a very hot early universe. The question is whether this can also be found experimentally.

2.3.2   Background Radiation

From the above considerations, it was clear that a hot early universe was expected. Hot means high temperature, a body of high temperature radiates. The higher the temperature, the more the maximum of its radiation moves into the short-wave range. We know this from turning on a hot plate. First the plate gets warm. You can't see heat, but you can feel it. Physically, heat is infrared radiation. If you wait a while, the plate starts to glow red. It has become hotter. Red radiation is visible to our eye, the wavelength of red light is shorter than that of infrared radiation. If you wait even longer, the plate no longer glows dark red, but bright red.

This simple experiment shows us that the emitted radiation must somehow be related to the temperature of a body. In physics, this is called Wien's law.

Digression Wien's law states that the product of the temperature of a body and the maximum wavelength at which it radiates is constant:

$$T\lambda_{max} = 2.89 \times 10^6 \text{nm} \tag{2.7}$$

If we take for example the 6000 K hot sun, then one finds: $\lambda_{max} = 489$ nm, this is in the green range of the visible spectrum" for a cool planet, e.g. earth, with T $= 300$ K amounts to $\lambda_{max} = 9630$ nm, thus in the infrared.

Let's apply this experiment to the evolution of the universe:

- Early universe: hot, so radiation was in the short wavelength range, UV, X-rays.
- Present universe: has expanded enormously, hence low temperature, hence radiation must be found at very long wavelengths.

Very often it happens that one looks for some effects and finds something quite different unexpected. Already around 1940 a radiation was predicted by G. Gamow, R. Alpher and R. Herman, which should come from the time of the early universe. Then in 1964 it was found as noise in antennas by A. Penzias and R. Wilson. The two scientists wanted to study the propagation of electromagnetic waves in the solar system, or rather, how their propagation changes when the sun, as a result of its activity, hurls enormous amounts of matter into interplanetary space. In the measurements, it was found that there is a noise that is received equally from all directions of the compass. The maximum of this radiation is in the radio range; the energy distribution (spectrum) of the background radiation corresponds to the temperature of a body of 2.7 K (K stands for Kelvin; in the Kelvin scale one counts away from absolute zero, which is -273.3 °C). Incidentally, Penzias and Wilson initially thought this radiation was a disturbance caused by bird droppings on the antenna. They were awarded the Nobel Prize in Physics for their discovery in 1978.

2.3.3 Background Radiation and Redshift

The background radiation comes from the early hot phase of the universe. Let's imagine a gas that consists mainly of hydrogen. Hydrogen is the most common and simplest element in the cosmos. It has a positively charged proton in the nucleus surrounded by a negatively charged electron in the shell. At high temperatures, however, the electrons are separated from the atom. Then the gas consists only of hydrogen ions. Radiation, i.e. electromagnetic waves, cannot penetrate such a gas, because scattering occurs at the free electrons. The gas becomes opaque. One can consider that at a temperature below 3000 K a recombination of the free electrons with the protons occurs, i.e. a transition from hydrogen ions to neutral hydrogen atoms. Thus the gas becomes transparent.

Digression We can now estimate when the temperature of the universe was about 3000 K and find the value of 400,000 years after the Big Bang. So the radiation we now observe as 2.7 K background radiation comes from a time when the universe was about 400,000 years old! We cannot look further back in time because then the universe becomes opaque. So the universe is observable from today to an age of about 400,000 years. In astrophysics, one often uses the redshift z, which was introduced earlier.

$$z = \frac{\Delta\lambda}{\lambda} = \frac{v}{c}. \tag{2.8}$$

The redshift of the background radiation is $z = 1,000$. The above formula would then result in a speed of $v = 1,000 \ c$, i.e. a thousand times the speed of light. Of course, there can be no velocities above the speed of light. We simply have to calculate with the relativistic Doppler formula:

$$\frac{\Delta\lambda}{\lambda} = \frac{\sqrt{1 + v/c}}{\sqrt{1 - v/c}} - 1. \tag{2.9}$$

This shows that when v gets close to the speed of light, the redshift z can become larger than 1. Let us assume that a radiation has a wavelength of 100 nm. This would be in the UV range, which is not visible to us (the range visible to the human eye is between 400 and 700 nm). A redshift of $z = 1,000$ means that we observe this radiation at a wavelength of.

$$\Delta\lambda/\lambda = z \qquad \Delta\lambda = z\lambda = 100 \times 1,000 = 100.000 \ \text{nm replace by,} \tag{2.10}$$

$$\lambda = \lambda_0 + \Delta\lambda = 100 + 100.000 = 100.100 \ \text{nm, replace by,} \tag{2.11}$$

i.e. we see the radiation at a wavelength of $10^5 \times 10^{-9} = 10^{-4}$ m, i.e. in the microwave range.

The 2.7 K background radiation is evidence for a hot big bang. However, we can only look back to about 400,000 years after the Big Bang, because then the universe becomes opaque to radiation.

2.3.4   Temperature Fluctuations in the Early Universe

Several satellite missions (COBE, WMAP, PLANCK) have been used to accurately measure the microwave background. This shows that there are minute variations in temperature (Fig. 2.10). However, the measured variations are very small, only about 0.001%. So on the one hand the background is remarkably uniform, but on the other hand there are still small fluctuations. These fluctuations may have acted as later condensation nuclei for galaxies. Fluctuations within early universe can be understood as vibrational motions. Here again is an example familiar from everyday life. If you pluck the string of a guitar, you will hear a certain tone. However, there are also numerous harmonics:

- Root note: a vibratory bulge along the string's entire length,
- first overtone: two vibrational bellies along the string,
- second overtone: three vibrational bellies.

This is outlined in Figs. 2.9 and 2.10.

Overall, a struck guitar string results in a combination of fundamental and harmonic oscillations. The early universe can be seen as an extremely hot compact plasma. In such a plasma there are always random disturbances. In physics, perturbations of an equilibrium state usually (actually almost always) lead to oscillations. These oscillations can also be measured in today's universe as temperature and wavelength fluctuations. The metric of the universe is given by its curvature and initial velocity. Therefore, one can infer these values from the analysis of the measured oscillations. The WMAP probe has measured over a thousand such harmonics.

The COBE satellite (Cosmic Background Explorer) was in operation from 1989 to 1993, its successors were WMAP (Wilkinson Microwave Anisotropy Probe), which measured from 2001 to 2010, and PLANCK (since 2009).

The results of the WMAP measurements are shown in Fig. 2.11. Red means warmer regions, blue means colder regions. The temperature fluctuations measured with this

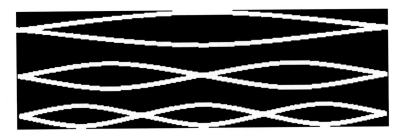

Fig. 2.9 Fundamental and first and second harmonics of a guitar string

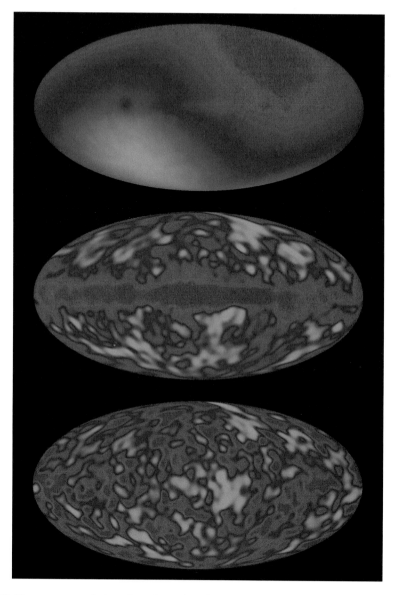

Fig. 2.10 Temperature variations from the early universe, measured with the COBE satellite. *At the top* the raw data are shown, influenced by the motion of the satellite, in *the middle one* sees the influence of the Milky Way, at *the bottom* the corrected final data. (NASA)

method are 5×10^{-5} K, i.e. less than measured with COBE. With WMAP (Fig. 2.11), the age of the universe was determined to be 13.75 billion years, with an uncertainty of 1%.

In the microwave background, there are tiny temperature differences from which galaxies later evolved.

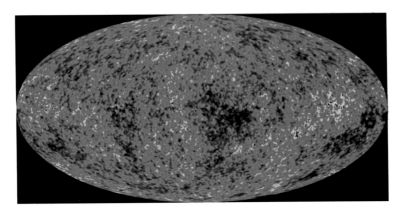

Fig. 2.11 Results of WMAP measurements showing temperature fluctuations in the early universe. (NASA)

2.3.5   Dark Matter

The observed temperature fluctuations cannot be explained from the distribution of visible matter alone. Assuming that about 22% is dark, non-radiating matter and that the visible luminous matter (luminous means here: matter radiates, no matter in which wavelengths) makes up about 4%, one can explain the distribution measured with WMAP. Dark matter thus exceeds the radiating matter known to us by at least a factor of 5! The PLANCK satellite provides even finer details of the background radiation. Figure 2.12 sketches what the PLANCK data tell us about the early universe. The satellite sees virtually through the solar system, galaxies, to the earliest galaxies, and then to a time when recombination was complete, that is, electrons united with protons; as we have seen, this process was complete at a temperature of 3000 K and the universe was about 400,000 years old.

So far we have discussed experimental observations that point to a big bang. There is one more observation that can only be explained by the big bang model.

Visible matter is not sufficient to explain the measured temperature fluctuations in the early universe. Therefore, there must also be dark, non-luminous matter.

Fig. 2.12 Sketch of how the satellite PLANCK sees to the first beginnings of the universe. Even deeper into the past nothing can be seen, the universe becomes opaque. (ESA)

2.4 The Origin of the Elements

2.4.1 Nuclear Fusion at the Beginning

Hydrogen is the most abundant element in the cosmos. The universe consists roughly of about 75% hydrogen and about 25% helium. All elements heavier than helium make up less than 1% of the mass. Nevertheless, these elements are important, without them there would be no solid planets and, of course, no life.

Where do these elements come from? If we look back into the past of the universe, we see that the composition 3/4 hydrogen, 1/4 helium remains almost constant. However, very old stars have an even lower content of metals, and under astrophysics anything heavier than helium is called a metal. This suggests that the ratio of hydrogen to helium was fixed in the early stages of the universe. This is called *primordial nuclear fusion*. The universe was still sufficiently hot and dense, helium atoms could be formed from four hydrogen atoms during a short period of time (within the first three minutes) by nuclear fusion. Helium consists of two protons and two neutrons, so two protons were converted to neutrons during fusion. Primordial nucleosynthesis limits the number of baryons ($\beta\alpha\rho\nu\varsigma$ means heavy; heavy particles consist of three quarks, e.g. neutrons, protons) in the universe. The non-uniform distribution of baryons in the universe is explained by the dark matter already discussed.

► At the origin of the universe, i.e. at the big bang, the elements hydrogen and helium were formed within the first 3 min, whereby the hydrogen was ionized because of the high temperatures.

2.4.2   Elements Heavier than Helium

A few minutes after the Big Bang, the temperature and density were no longer high enough to allow nuclear fusion. Where then do the other elements, e.g. oxygen or carbon, come from?

These elements could not have been created during the Big Bang because the universe expanded very rapidly and cooled as a result. All elements heavier than helium were created inside stars by nuclear fusion. It is possible that the first stars to form in the early universe were much more massive than stars as we know them today. In discussing stellar evolution, we will see that the evolution or lifetime of a star is determined by only one parameter: its *mass*.

► The more massive the stars, the shorter their lifetimes.

The early stars (called Population III stars) evolved very rapidly (within less than about one million years) and then exploded, enriching the universe with elements heavier than helium. So, in that sense, our Earth, like the other Earth-like planets, is made of stardust.

The formation of the elements in the universe happened and still happens in two phases:

- During the first three minutes: primordial nuclear fusion, hydrogen (protons) becomes helium (slightly less than 25%).
- Already in the first stars, heavier elements are then formed by nuclear fusion processes in the interior. This process is still going on today. Slowly the universe is enriched with heavier elements.

► All elements heavier than helium were formed by fusion in the stellar centers. So we are made of *stardust*.

2.5   The Early Universe

After examining experimental evidence that supports a big bang model (big bang theory), we still examine the early universe.

2.5.1 The Superforce

In the first chapter we discussed the four basic forces that govern the universe today: Gravity, Electromagnetic Interaction, Strong Interaction and Weak Interaction. The structures of the universe today are governed by the weakest force, gravity. However, if we go back to the early stages of the universe, the energy density, and therefore the temperature, becomes greater and greater. At higher and higher energies, a unification of the forces of nature occurs. If we go further and further back in history to the beginning, we first find a unification of the weak interaction with the electromagnetic interaction. This has already been verified experimentally in large accelerators. We then speak of an *electroweak interaction*. In 1979 Glashow, Salam and Weinberg received the Nobel Prize for their theory of the electromagnetic and weak interaction, and in 1984 Rubbia and van der Meer discovered the W and Z particles predicted by this model.

When the universe was about one millionth of a second old, $t \sim 10^{-6}$ s, the temperature was 10^{13} K, and a unification of the electromagnetic interaction with the weak interaction occurred. At even higher temperatures, when the universe was 10^{-35} s old, the electroweak interaction united with the strong interaction. This is referred to as GUT. GUT stands for Grand Unified Theory. At even higher energy, all four forces of nature known today were unified into one super force.

The splitting of the forces can be thought of as a kind of *symmetry breaking*. Here is an analogy. Let us take the different phase states of water as an example: ice, liquid, gaseous. Ice has the highest symmetry (order), in the liquid state the water molecules are much freer to move and in the gaseous state this freedom is greatest. If you cool water to below zero degrees, crystallization does not start immediately, the water is supercooled. Only at temperatures below zero degrees does solidification begin spontaneously and energy is released.

The superforce has the highest symmetry, no directions are excellent. As an analogue we would have water vapor. With cooling (=expansion of the universe) phase transitions occur.

The four basic forces known today developed by splitting off (symmetry breaking) in the early universe from a single super force.

2.5.2 The Inflationary Universe

The time 10^{-35} s after the big bang was a special one. As already described above, a symmetry breaking occurred, the strong interaction separated from the electroweak interaction. The universe was filled with energy, also called *vacuum energy*. This made gravity repulsive for an extremely short time and the universe expanded by a factor of 10^{30}. This rapid expansion of the early universe is called the *inflationary phase*. Once the phase transition or symmetry breaking was complete, the evolution of the universe resumed

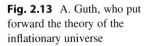

Fig. 2.13 A. Guth, who put
forward the theory of the
inflationary universe

normally. After the inflationary phase was completed, the universe had an extension of
about 1 m.

The theory of the inflationary universe, developed by A. Guth (Fig. 2.13), follows on the
one hand from the concept of symmetry breaking, on the other hand it can also explain
some experimental findings:

- Horizon problem: Consider two opposite points in the sky. These can never have been
  physically related because of their great distance. However, an inflationary expansion
  explains this, i.e. before the inflationary phase there was very well a connection.
- Flatness problem: The universe appears extremely flat despite its mass.
- Magnetic monopoles: Magnetic fields are characterized by the fact that there is always a
  north and south pole. Magnetic monopoles, however, could have originated in the early
  phase, but have been widely dispersed by the inflationary phase and are therefore
  practically unobservable today.
- Density fluctuations: Due to the inflationary expansion, small density fluctuations
  developed or intensified, which then later served as condensation nuclei for matter,
  i.e. galaxy clusters/galaxies.

Imagine you are on a celestial body that is spherical. The larger the sphere, the less one will
perceive something of the curvature on small scales.

In Fig. 2.14, the evolution of the universe is shown schematically, as is symmetry
breaking (splitting of forces).

▸ The inflationary phase in the early universe is needed to explain why the universe
appears to us today to be quite flat.

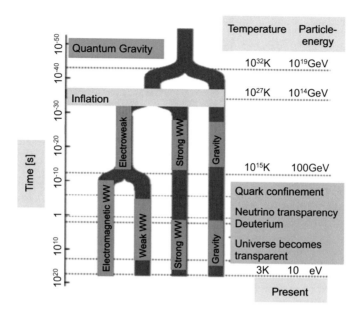

Fig. 2.14 Evolution of the universe, symmetry breaking, splitting of forces

2.6   Time Scale

We scale the evolutionary history of the universe to 1 year. Thus, 1 year means the entire time span that has passed since the Big Bang. Then we find the values given in Table 2.1.

This table also says something about the future. At the "end of the year" we have arrived at the present time. As our Sun slowly gets hotter, the temperature on Earth begins to increase over the next 100 million years (of course, that doesn't explain the current global warming) and on our time scale, Earth becomes uninhabitable by mid-January. Finally, our sun is expanding into a red giant across Earth's orbit. We are also due for a collision with the Andromeda Galaxy.

Table 2.1 Evolutionary history of the universe scaled to one year

Time	Event
Jan 10^h0^m	Big bang, origin of H, He
Jan 10^h14^m	Decoupling of radiation and matter
Jan. 5.	First stars and black holes, C, N, O, …
Jan. 16.	Oldest known galaxy, quasar
9 Sep.	Origin of the solar system and the earth
28 Sep.	First life on earth (cyanobacteria)
Dec. 16–19.	Vertebrate fossils, plants
Dec. 20–24	Forest, fish, reptiles
Dec. 25.	Mammals
28 Dec.	Extinction of the dinosaurs
31 Dec 20^h00^m	First people
31 Dec 23^h55^m	Neanderthal
31 Dec $23^h55^m56^s$	The year 0
12 Jan	Earth gets too hot
April 7	Sun becomes a red giant
April 16	Collision milky way with Andromeda galaxy

2.7   The Future of the Universe

2.7.1   Mass and Energy

Let's start again with an illustrative example. Let's try to throw a stone upwards. Without going into the exact formulas, it is clear that the height reached by the stone depends on the initial velocity. The greater the initial velocity, the greater the height reached.

Digression One can also imagine that at a certain speed the kinetic energy given to the stone becomes so great that it can leave the area of attraction of the earth or can first circle around the earth in an orbit – we know this from satellites, of course. This orbital velocity v_k is:

$$v_k = \sqrt{GM/R}, \tag{2.12}$$

where R *is* the earth's radius (6371 km), M the earth's mass (5.97×10^{24} kg), G the gravitational constant (6.67×10^{-11}). A satellite at a height of 300 km above the earth's surface then orbits the earth in about 90 min. The speed to finally fly away from the earth is then

$$v_{\text{entw}} = v_k \sqrt{2}. \tag{2.13}$$

If we use the values for the earth, we find that if we bring a stone or a rocket with a speed of 11.2km/s upwards, they will escape from the earth.

The universe was created 13.7 billion years ago by the Big Bang. It has been expanding ever since. The question is, how long does this expansion last? In principle, there are three possibilities:

- Expansion lasts forever; that would be the case for an open universe.
- Expansion is slowly going towards zero, more precisely, for $t \to \infty$; this would be a borderline case.
- Expansion ends at some point; then contraction occurs again. That would then be a closed universe.

What could stop the expansion? Similar to the example of the stone thrown upwards, where it depends on the mass of the earth how high the stone flies or whether it flies away from the earth, with the universe it depends on its total mass whether it is open or closed. Therefore, there must be some kind of critical matter density of the universe, ρ_{crit} . The universe is

- open if the matter density $\rho < \rho_{\text{crit}}$,
- Expansion stops at infinity when $\rho = \rho_{\text{crit}}$,
- Expansion turns into contraction after a certain time when $\rho > \rho_{\text{crit}}$.

▶ Therefore, we need to know the critical density of matter and compare it to the measured density in order to answer the question of whether the universe will expand forever or collapse again at some point.

The critical matter density determines the expansion, and the expansion is in turn related to the Hubble constant, therefore the critical matter density is proportional to the Hubble constant.

Digression

$$\boxed{\rho_{\text{crit}} = \frac{3H_0^2}{8\pi G}.} \tag{2.14}$$

Here H_0 is the present value of the Hubble constant.

One also specifies a total energy K: This parameter can be

- $K = +1$ total positive energy, space curvature $k < 0$, hyperbolic case $\to$ expansion takes forever;
- $K = 0$, total energy $=0$, space curvature $k = 0 \to$ expansion stops after infinite time, parabolic case;

- $K = -1$, negative total energy, space curvature $k > 0$, spherical case $\rightarrow$ expansion eventually turns into collapse.

One can introduce a density parameter that describes the ratio of the measured matter density ρ to the critical density:

Digression
$$\Omega_m = \rho/\rho_{crit} \tag{2.15}$$

and can show that

$$\Omega = 1 + \frac{Kc^2}{HR^2}. \tag{2.16}$$

The value determined for the critical density of matter is

$$\rho_{crit} = 9 \times 10^{-27} \ \mathrm{kg/m^3}. \tag{2.17}$$

To this critical matter density one must count everything what is subject to gravity. We must therefore include dark matter. Then one finds:

$$\Omega_m = 0.3. \tag{2.18}$$

What does this mean? The density of matter in the universe is only about 30% of the critical density, so this would mean expansion for all eternity.

In Fig. 2.15 the three limiting cases are shown again. In the *y-axis* the expansion of the universe is given, which is equivalent to its expansion. In the *x-axis* the time is plotted. For $\Omega < 1$ the expansion lasts forever.

Fig. 2.15 The parameter Ω, i.e. the ratio of matter density to critical matter density, determines the future of the universe

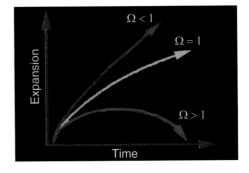

2.7.2 The Non-empty Vacuum

According to quantum field theory, there is no actual vacuum, but virtual particles are continuously created and destroyed again. This leads to a non-vanishing energy-momentum tensor of the vacuum. In short: There is energy in the vacuum. The *vacuum energy density* $\Omega_L = 0.7$ therefore corresponds to 70% of the critical energy density.

This vacuum energy density leads to an accelerated expansion of the universe. Imagine an ordinary explosion. The pieces fly apart. One would expect the expansion to slow down over time. So it used to be faster than it is now. Exactly the opposite has been measured for the universe! The universe is expanding at an accelerated rate. Since acceleration requires energy, the energy required for this has also been called *dark energy*. What exactly dark energy is, we do not know. But how do we know at all that the universe expands accelerated?

▸ The universe is expanding at an accelerated rate, now faster than before, this requires dark energy.

As we mentioned, the deeper you look into the cosmos, the more you look into the past. So distant galaxies are much younger than nearby galaxies. If we measure the redshift of distant galaxies and compare them to closer galaxies, we get the change in expansion rate. The result shows: The universe is expanding faster today than it used to!

Digression The Casimir effect is evidence of vacuum fluctuations. Consider two parallel plates (Fig. 2.16). As shown in the figure, a force acts on them in a vacuum, pushing the plates together. Outside the plates there exists a continuum of virtual particles, inside the plates there can be only a discrete number of particles. Corresponding to the particles are the so-called De Broglie wavelengths. Any particle with momentum p has a wavelength of

Fig. 2.16 The Casimir effect, evidence for the existence of vacuum fluctuations. (Wikimedia Commons; cc-by-sa 3.0)

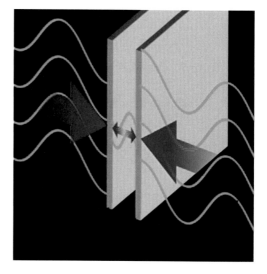

$$\lambda = h/p, \tag{2.19}$$

Where p is the relativistic momentum: $p = mv/\sqrt{1 - (v/c)^2}$.

2.7.3   Matter and Space

According to Einstein's general theory of relativity, there is a connection between matter and space. Matter determines the curvature of space, or more precisely of space-time. This can be seen in the field equations of *general relativity* . The greater the mass, the greater the curvature of space. This theory could be proved experimentally. One of the first confirmations was the observation of the *deflection of light*. During a total solar eclipse the sun is eclipsed by the moon as seen from the earth. Thus, stars can be observed even during the day. If one measures the position of stars near the edge of the sun and compares this measured position with the measured position when the sun is not near these stars, then one finds a small deviation. This deviation can be explained by the deflection of the light, as shown in Fig. 2.17. In 1919, an expedition funded by the Royal Society was launched to make this measurement. Near the edge of the Sun the deflection of light is only 1.75 arc seconds, so very accurate measurements are needed to find this effect.

The deflection of light, like many other effects of general relativity, depends on the ratio of the Schwarzschild radius to the radius of the object. The Schwarzschild radius is calculated as follows: consider what an object would have to be like in order for its escape velocity to be equal to the speed of light, *c:*

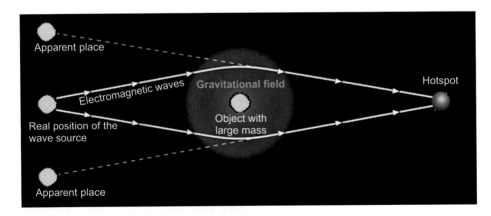

Fig. 2.17 Due to the deflection of light in the presence of masses, the position of a star located behind an object with a large mass appears shifted; several images of the star are observed at the focal point (on Earth). (www.scififorum.de)

Digression

$$c = \sqrt{2GM/R}, \tag{2.20}$$

this gives the *Schwarzschild radius*:

$$R_S = \frac{2GM}{c^2}, \tag{2.21}$$

Let's put the values for the sun into the formula: The mass of the sun is $M_\odot = 2 \times 10^{30}$ kg, then we get 3 km for the Schwarzschild radius of the sun. So if the Sun were compressed from its present radius of about 700,000 km to 3 km, it would be a black hole, since not even radiation would be able to leave its surface. Our earth would have to be compressed to about 1 cm to become a black hole.

Digression We can now consider the universe as a whole as mass. Of course, this also results in a curvature of space (space-time). In the theory of relativity one always speaks of space-time. An event in the universe has three spatial coordinates (x, y, z) and one temporal coordinate (ct). The properties of space can be specified by looking at the distance ds^2 between two events in space-time:

$$ds^2 = c^2 dt^2 - dx^2 - dy^2 - dz^2. \tag{2.22}$$

So the mass of the universe bends space-time. Mass determines space. It is therefore pointless to ask what lies outside the universe. Space-time is determined by the mass contained within the universe. Similarly, it is pointless to ask what was before the Big Bang. Space-time was determined and defined by the big bang.

▸ Space and matter are linked by general relativity.

As already mentioned, Einstein expressed this in his famous field equations:

- The left side of the field equations contains the properties of space-time, i.e. the curvature of space.
- The right-hand side of the field equations contains the distribution of matter in space.
- Therefore we see: Matter and space-time curvature are connected.
- If the universe is static, as was thought at the time the field equations were published, then the universe would have to collapse because the masses attract each other. To maintain a static universe, Einstein introduced the Cosmological Constant. When Hubble proved the expansion of the universe, this constant, called Λ, was no longer necessary. Today, Λ is used again to bring the accelerated expansion of the universe into the equations.

2.7.4 Energy and Mass

Almost everyone knows Einstein's most famous formula. It represents a simple relationship between the energy and the mass of an object:

Fig. 2.18 Sketch showing the generation of an electron-positron pair in a bubble chamber

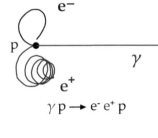

$$E = mc^2. \tag{2.23}$$

From this we can see: With sufficiently high energy, we can produce particles. Conversely, some mass is lost during nuclear fusion, and the difference is radiated as energy. As an example, consider the fusion of four hydrogen atoms into helium. The helium atom is slightly lighter (less than 1%) than the sum of the four hydrogen atoms, and this difference in mass is radiated as energy.

Pair generation involves the creation of particles at high energies. This was first demonstrated in 1933, as electron-positron pair production (the positron is the antiparticle to the electron, it has the same properties as the electron, but has the opposite charge, i.e. +e) (Fig. 2.18).

▸ At high energies particles are created, this was the case in the early phase of the cosmos.

So the further back we go in the history of the universe, the higher temperatures we find and heavier and heavier particles could be created.

The history of the universe is sketched in Fig. 2.19.

Thus we have set out the history of the universe in the so-called standard model of cosmology or particle physics. The further back we go, the higher the energies become. Attempts are being made to reproduce some of the processes that took place in the early universe using modern accelerator facilities (e.g. CERN).

2.7.5 Big Bang and Planck Era

If we calculate the expansion backwards, we come to a singularity, the universe becomes a point. But our current physics already fails shortly before that. We can calculate up to about 10^{-43} s after the big bang, which is the so-called Planck time. At that time the universe has a length of 10^{-35} m and a temperature of 10^{32} K, the density was 10^{94} g/cm^3 .The time before this is called the Planck era. There could have been a quantum vacuum with perfect symmetry during this time, which had infinite dimensions. In such a quantum vacuum, spontaneous symmetry breaking could have occurred. So is our universe the result of spontaneous symmetry breaking?

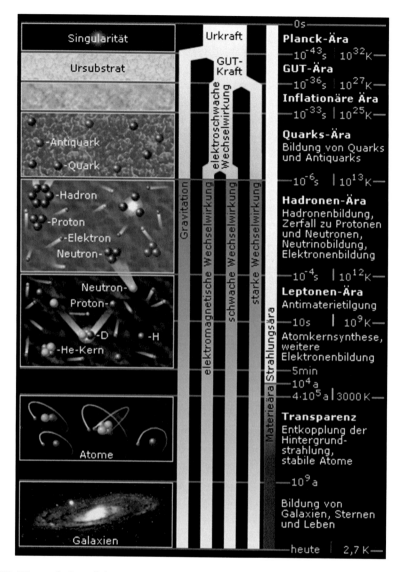

Fig. 2.19 The evolution of the universe. (Source: http://sfrd.u-f-p.net/2010/03/der-urknall/)

2.8 The Universe and Particles

2.8.1 Boson Era

Bosons are particles that transmit the forces between the particles of matter, the fermions. The Higgs particle is named after P. Higgs. It is electrically neutral and has spin 0. According to Higgs, all elementary particles acquire their mass only through interaction with the ubiquitous Higgs field. The Higgs boson belonging to this field was found in 2012 during experiments at CERN. The Standard Model of theoretical elementary particle physics recognizes four types of particles:

- Quark's
- Leptons
- gauge bosons – they mediate interactions, i.e. forces
- Higgs Field

In physics there is the second quantization, the opposition between particle and field is cancelled.
   ► A particle is an excited state of the corresponding field.
   The Higgs boson is a quantum mechanical excitation of the Higgs field. This particle can be detected at high energies. Imagine a guitar string. This is excited to a vibration. The vibration is heard as a tone.

- Guitar string: corresponds to the Higgs field,
- Tone: corresponds to the Higgs particle.

2.8.2   Quark Era

At the age of 10^{-33} s the universe has cooled down to 10^{25} K. Before, the energetic photons formed the heavy X- and Y-bosons, now the energy is no longer sufficient to form this, and quarks/antiquarks are formed. In addition, leptons such as electrons and neutrinos already exist. However, due to the high energy, there are no neutrons and protons yet. So no matter known to us. The matter at this time can be described as a *quark-gluon plasma*. Already in the year 2000, such a quark-gluon plasma could be generated and indirectly detected at the Swiss CERN for an extremely short time.
   When the universe has reached an age of 10^{-12} s, the temperature has dropped to 10^{16} K. It comes to the splitting off of the electroweak force into the weak force and electromagnetic force. So from this point on, all four forces of nature known to us existed in the universe.

2.8.3   Hadron Era

As the universe ages, it gets colder and colder. When it was 10^{-6} s old, the quarks formed the hadrons, of which essentially only the protons and neutrons survived. We have discussed the creation of particles and found that particles/antiparticle pairs are always created and then destroyed. Then why are there particles in the universe at all if everything immediately re-radiates due to pair annihilation? The reason lies in a very small asymmetry. There were 1,000,000,001 more particles than antiparticles. Therefore there is a cosmos with matter.

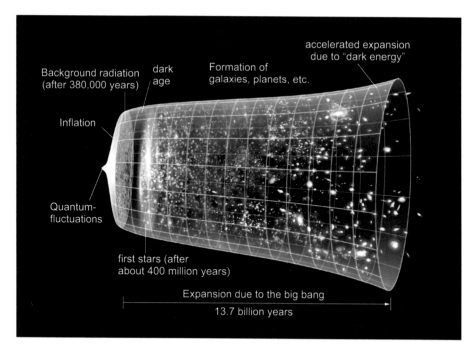

Fig. 2.20 The most important points in time in the evolutionary history of the universe. (NASA)

2.8.4 Lepton Era

At the age of 10^{-4} s the temperature of the universe is still 10^{12} K. The energy of the photons is only enough to form lepton pairs. When the universe is 1 s old and the temperature is 10^{10} K, electron-positron pair annihilation occurs, but again there is a small asymmetry and the electrons survive.

2.8.5 The Universe Becomes Transparent

Up to an age of 379,000 years, the universe remains opaque. The photons scatter by the free electrons. The electrons gain energy in the process. Now the temperature of the universe has cooled down to 3000 K. The free electrons are captured by the protons and neutral hydrogen atoms are formed. This is called recombination. From this point on the universe became transparent.

In Fig. 2.20, the evolution of the universe is shown again schematically.

The World of Planets

3

In this section we cover the planets of the solar system. We define eight major planets today: Mercury, Venus, Earth, Mars, Jupiter, Saturn, Uranus, and Neptune. Pluto is no longer counted among the major planets, since many objects have now been found in the solar system that are similar in size to Pluto.

In general, the large planets of the solar system are divided into two classes. The Earth-like or terrestrial planets have a solid surface, the gas planets Jupiter, Saturn, Uranus and Neptune are much larger and consist mainly of gas and a solid core.

A size comparison of the planets with the Sun is given in Fig. 3.1.

After reading this chapter you will have an overview of

- the structure of the planetary system,
- can compare planets with each other,
- understand the meaning of comparative planetary research,
- understand how, for example, by comparing the Earth's atmosphere with the atmosphere of Venus, information can be obtained about the consequences of global warming of the Earth,
- learn about the bizarre worlds of the planetary moons.

3.1 General Properties of the Planets

3.1.1 Mass and Radius

One of the most important physical state variables of planets is their mass. This can be determined relatively easily from Kepler's third law, if the planet is orbited by a moon.

© The Author(s), under exclusive license to Springer-Verlag GmbH, DE, part of
Springer Nature 2023
A. Hanslmeier, *Fascination Astronomy*,
https://doi.org/10.1007/978-3-662-66020-1_3

Fig. 3.1 Sun with the planets. Earth is the third planet as seen from the Sun. (NASA/Wikimedia Commons)

Digression We explain this by the example of the determination of the mass of the earth. We use the exact form of Kepler's third law for this.

The moon orbits the earth in about 27 days, this is its orbital period in relation to its position in the sky. This period is also called the sidereal month. The distance Earth-Moon is on average 384,400 km. Thus one knows the variables $a = 384.400$ km, $T = 27$ days. The mass m_2 of the moon can be neglected in the formula and one gets the mass of the earth m_1 from the general form of the third Kepler law ($G = 6.67 \times 10^{-11}$ m^{3}kg^{-1}s^{-2}, gravitational constant):

$$\frac{a^3}{T^2} = \frac{G}{4\pi^2}(m_1 + m_2). \tag{3.1}$$

By the way: The moon mass is only 1/81 of the earth mass.

A major problem in mass determination has long been that of Pluto, now classified as a dwarf planet. Pluto is relatively small, it exerts little perturbation on the masses of the other planets (mainly Neptune's orbit). It was not until moons were discovered around Pluto that its mass could be accurately determined. The planet with the largest mass in the solar system is Jupiter with more than 300 Earth masses. Compared to the mass of the sun, however, its mass is also small, only 1/1,000 solar mass.

The radius or diameter of a planet can be determined if one knows its distance as well as the apparent diameter of the planet disk. A simple trigonometric calculation gives the diameter. The diameter of Jupiter is about 10 times that of Earth. If you line up Jupiter 10 times, you get the diameter of the sun.

▸ Kepler's Third Law allows the determination of planetary masses from the orbital motion.

3.1.2   Distances

There are two ways to determine the distances of the planets. One method is to measure the *parallax*. If one looks at a close planet (e.g. Venus) from two points as far away as possible from the earth, then one can determine the parallax of Venus as shown in Fig. 3.2. Since one knows the distance d of the two measuring points on Earth, the distance of the planet follows immediately. However, once one distance in the solar system is calculated, all others are known directly from Kepler's Third Law, which states that the ratio of the cubes of the distances to the squares of the orbital periods is constant; thus, if a_1, T_1 denote the distance and orbital period, respectively, *of the* first planet, and a_2, T_2 that of the second, then:

$$\frac{a_1^3}{T_1^2} = \frac{a_2^3}{T_2^2}. \tag{3.2}$$

For investigations in the solar system, the average distance from the Earth to the Sun is usually used as the unit of distance. This is called the *astronomical unit,* abbreviated AU (Astronomical Unit).

One astronomical unit 1 AU $=$ 150,000,000 km.

3.1.3   Temperatures, Atmospheres

The temperature of a planet can be easily estimated by calculating the radiant power that the surface of the planet receives from the sun. On Earth, under ideal conditions, we receive about 1.36 kW per m^2. However, this is only true if the sun's rays are incident perpendicularly and if nothing is absorbed in the earth's atmosphere. Since the irradiation decreases with the square of the distance, on a planet which is twice as far away as the earth from the sun, we receive only 1/4 of the amount of 1.36 kW per m^2.

Apart from Mercury, which has only an extremely thin atmosphere, all planets have atmospheres. These atmospheres change the surface temperatures of the planets tremendously. Our Earth's atmosphere contains natural greenhouse gases such as water vapor, carbon dioxide (CO_2), and without the natural greenhouse effect, the Earth's surface would be about 30° cooler. The actual measured global surface temperature of the Earth is about 14°, without the greenhouse effect it would be minus 16°, life would probably be

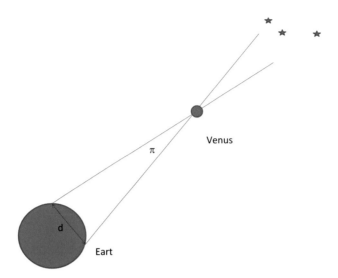

Fig. 3.2 To determine the Venus parallax π. One observes at two points separated by the distance d. Since the angles are very small, the simple relation $\pi = a/d$ applies, where a is the distance of Venus from the Earth

impossible on Earth. The atmosphere also protects planets from cooling during the night. Our moon has no atmosphere. Where the sun shines, the temperature is over $120°$ where there is shade, it is $-180°$ on the lunar surface.

The composition of the planet's atmosphere can be determined from analysis of the spectrum. The planets reflect sunlight, but one also observes dark absorption lines in the spectrum that originate from the planetary atmospheres themselves.

▶ Planetary atmospheres protect planets from excessive temperature contrasts. Without the natural greenhouse effect (greenhouse gases that exist independently of anthropogenic emissions) our earth would be uninhabitable.

3.1.4   How Do We See Planets in the Sky?

In ancient times, five planets were already known: Mercury, Venus, Mars, Jupiter and Saturn. These objects plus the sun and moon also make up the seven-day week. Monday is the day of the Moon, Tuesday (French Mardi) the day of Mars, Wednesday (French Mercredi) the day of Mercury, Thursday (French Jeudi) the day of Jupiter, Friday (French Vendredi) the day of Venus, Saturday (English Satuday) the day of Saturn and Sunday the day of the Sun.

The orbit of the moon around the earth roughly defines the month. Here, however, one must distinguish between the synodic and sidereal month. Synodic month means from a certain lunar phase to the next occurrence of this phase. So if there is a full moon today,

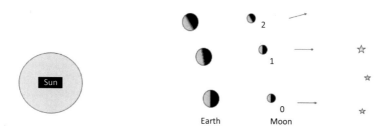

Fig. 3.3 The difference between sidereal and synodic orbits of the Moon. After a sidereal orbit (Moon position 0 after 1) the Moon is at the same position in the sky, but only after a synodic orbit (Moon position 0 after 2) it shows the same phase

then after 29.5 days or a synodic month, the moon will be full again. The sidereal month is only about 27 days, which is how long it takes for the moon to return to the same position among the stars. Since the earth moves around the sun, the synodic month is longer than the sidereal month.

This is illustrated in Fig. 3.3. At position 0, the moon is full. At position 1, the moon has the same position in the sky relative to the stars, but only at position 2 is the moon full again.

With the position of the planets relative to the earth one distinguishes between planets, whose courses lie within the earth course and/or whose courses lie outside of the earth course.

The two planets Mercury and Venus, also called inner planets, are never visible throughout the night, but only reach a maximum angular distance east or west of the Sun. Therefore, around the time of their maximum *elongation,* they are seen in the morning sky when they are west of the Sun, or in the evening sky when they are east of the Sun. They are closest to the Earth during their inferior conjunction, when they are between the Sun and Earth, and on rare occasions, when the orbital planes coincide, a *transit* occurs and the dark disk of the planet is seen passing in front of the solar disk. The last transit of Venus occurred on June 6, 2012 (Fig. 3.4). If you didn't observe this one, there won't be another opportunity until December 11, 2117. The next Mercury transits are on November 13, 2032, and November 7, 2039, but Mercury transits can only be observed with a telescope (be sure to use protective film in front of the telescope!!!). Apart from the transits, Mercury and Venus are unobservable with the Sun in the daytime sky around the time of their lower conjunction. At their upper conjunction, you also can't see the planets, they are farthest from Earth then. The inner planets also show phases, which can already be observed with smaller telescopes, especially with Venus. A few weeks before or after its lower or inferior conjunction Venus can only be seen as a narrow relatively large crescent in the telescope. A few weeks before or after the upper conjunction Venus can be seen almost fully illuminated, but clearly smaller in diameter due to the greater distance to Earth (see Fig. 3.5).

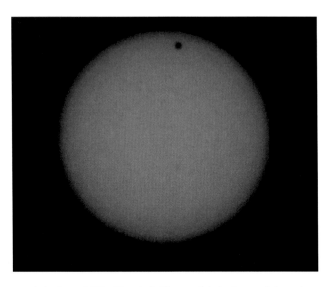

Fig. 3.4 Venus transit in June 2012. The dark Venus disk in front of the solar surface and some sunspots are clearly visible. (A. Hanslmeier, Private Observatory)

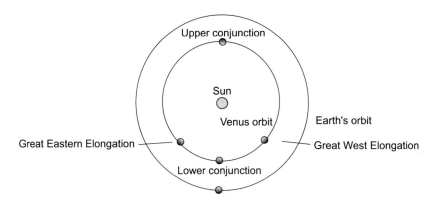

Fig. 3.5 The positions of an inner planet

The outer planets Mars, Jupiter, Saturn, Uranus and Neptune show the following special positions in relation to the Earth (see Fig. 3.6): At the time of their *opposition* they are opposite the Sun, they are then observable throughout the night, i.e. they rise when the Sun sets and set when the Sun rises. At the time of conjunction they are unobservable with the Sun in the daytime sky. Table 3.1 shows the opposition positions of the outer planets.

One feature that makes bright planets easily distinguishable from other bright stars is that planets, unlike stars, hardly blink at all. Because of their proximity, we on Earth get a beam of radiation from the planets, so they don't appear exactly point-like, and therefore the variations in temperature and pressure in Earth's atmosphere that cause stars to blink have less of an effect. Venus is easy to spot. Unless it is a few weeks before or after its

Fig. 3.6 The positions of an outer planet using Mars as an example

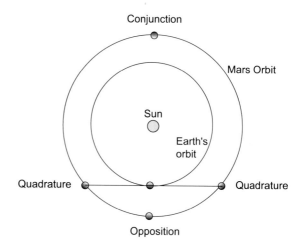

Table 3.1 Oppositions of the planets Mars, Jupiter, Saturn, Uranus and Neptune (e.g. 20.8 mean opoosition on August 20th)

Planet	2021	2022	2023	2024	2025	2026	2027	2028
Mars	–	8.12	–	–	16.1	–	19.2	–
Jupiter	20.8	26.9	3.11	7.12	–	10.1	11.2	12.3
Saturn	2.8	14.8	27.8	8.9	21.9	4.10	18.10	30.10
Uranus	5.11	9.11	13.11	17.11	21.11	25.11	30.11	3.12
Neptune	14.9	16.9	19.9	21.9	23.9	26.9	28.9	30.9

upper conjunction, it can be seen as a morning or evening star and is always the brightest object in the night sky after the Moon. Jupiter is the second brightest planet after Venus. At the time of its opposition it can be seen throughout the night. Mars has a reddish glow, especially when it is close to Earth (i.e. around its opposition). Its orbit around the Sun is relatively highly eccentric, so its minimum Earth distance varies from opposition to opposition from a minimum of 54 million to a maximum of more than 100 million km away. At extremely close Mars oppositions, it can even become brighter than Jupiter (this will be the case at its opposition in 2035!). Saturn is not as conspicuous as the other planets, but as bright as the brightest stars. The other outer planets, Uranus and Neptune, are invisible to the naked eye.

Some properties of the planets are given in Table 3.2.

▸ Planets within the Earth's orbit (Venus, Mercury) can be seen either as morning or evening stars. Planets outside the Earth's orbit are visible around their opposition throughout the night.

Table 3.2 The large planets. D equatorial diameter, v_e escape velocity

Planet	D [km]	M [M_{Erde}]	ρ [g/cm³]	Gravity Earth = 1	v_e [km/s]
Mercury	4878	0.055	5.43	0.4	4.25
Venus	12,104	0.815	5.24	0.9	10.4
Earth	12,756	1.0	5.52	1.0	11.2
Mars	6794	0.107	3.93	0.4	5.02
Jupiter	142,796	317.8	1.33	2.4	57.6
Saturn	120,000	95.15	0.70	0.9	33.4
Uranus	50,800	14.56	1.27	0.9	20.6
Neptune	48,600	17.20	1.71	1.2	23.7

3.2   The Earth-Like Planets

Mercury, Venus, Earth and Mars are called Earth-like or terrestrial planets. They have a solid surface and range in diameter from about 5,000 to 13,000 km.

3.2.1   Earth

Structure We begin with a brief description of the Earth as a planet. Seen from the Sun, it is the third planet. The surface of the Earth is relatively young, it is subject to constant changes. The average density of the Earth is 5,500 kg/m³, it increases towards the interior of the Earth. Near the center the density is about 13,000 kg/m³ . From the evaluation of the propagation of earthquake waves, we know relatively well about the Earth's interior. It has a shell-like in structure, the Earth's crust is up to 35 km deep above the continents, but only up to 5 km deep below the ocean floors. The earth's crust floats in large blocks above the earth's mantle, which becomes more and more liquid towards the inside. The plates move against each other (tectonics) and earthquakes can occur where there is tension due to the shifting of the plates. The earth's core is liquid on the outside, but solid on the inside and consists mainly of the metals iron and nickel.

The structure of the Earth with convection currents in the mantle leading to the continental drifts is sketched in Fig. 3.7.

The distribution of the continental plates has an important influence on the ocean currents and these in turn on the Earth's climate. 225 million years ago, the supercontinent Pangaea broke up into today's continents. Even today, continental drift amounts to a few centimetres per year and can be precisely determined using satellites. Why is the Earth's interior warm, what drives plate tectonics? The main reason is the radioactive decay of the elements. If we drill a hole in the Earth's crust, the temperature increases by about 30 K per

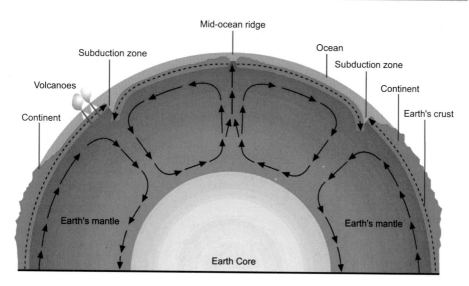

Fig. 3.7 Structure of the Earth with convection currents in the mantle. (© dpa/picture alliance)

km drilled. This shows the importance of using geothermal energy (e.g. for geothermal heat pumps). In the earth's core, the temperature is slightly less than 10,000 K.

Magnetic Field The liquid outer core of the Earth also explains the Earth's magnetic field, which near the Earth appears like a dipole field (dipole means two poles, i.e. a north and a south pole). However, the magnetic north pole is near the geographic south pole and conversely the magnetic south pole is in Canada, relatively close to the geographic north pole. Further away from the Earth's surface at a distance of a few Earth radii, the shape of the Earth's magnetic field is determined by the incoming solar wind. The solar wind consists of charged particles that are ejected from the Sun. In so-called coronal mass ejections (CMEs), large amounts of plasma are thrown away from the Sun and the Earth's magnetic field is additionally compressed on the side facing the Sun as shown in Fig. 3.8. Note that in this figure the scales are not true to nature. The Earth's magnetic field is very important to us. It protects us on the Earth's surface from the charged particles of the solar wind. The surface of the moon is exposed to this bombardment of charged particles.

Atmosphere The earth has an atmosphere, without which life would not be possible and which is divided into the following layers:

- Troposphere: extends to an altitude of about 12 km (at the equator); practically all weather events take place here. In the troposphere, air heated by the ground rises, then cools and clouds are formed. The temperature drops in the troposphere to about −50°. The boundary with the nearest stratosphere is called the stratopause.

Fig. 3.8 Sketch showing how solar wind particles affect the Earth's magnetic field. The particles cannot penetrate perpendicular to the field lines. However, some make it to the poles of the Earth's magnetic field and cause auroras. (NASA)

- Stratosphere: It reaches up to an altitude of about 50 km. There is an increase in temperature as a result of the absorption of short-wave UV radiation from the sun in the ozone layer. Ozone O_3 is formed from the combination of oxygen molecules. Without the ozone layer, life on the earth's surface would not be possible, as the short-wave UV radiation would penetrate to the earth's surface.
- Mesosphere: After the stratopause the mesosphere begins, where the temperature decreases again and at about 90 km altitude it is coldest with $-90°$.
- Ionosphere, exosphere: The temperature increases strongly here due to the absorption of short-wave radiation. Temperatures above 1,000 K are reached. However, it must be said that these layers are very thin and that we are dealing with kinetic temperatures. As we know from thermodynamics, the temperature is nothing else than a measure for the kinetic energy of the gas particles (which is given by $\frac{1}{2}mv^2$, m is the mass of the particle, v the velocity). Due to the high energy of short-wave radiation, ionization occurs in the ionosphere.

The variation of the temperature in the Earth's atmosphere is shown as a sketch in Fig. 3.9.

Earth's atmosphere is only transparent in two wavelength ranges, referred to as the two windows: the optical window goes from 290 nm (near UV) to about 1 µm (near infrared). The radio window ranges from 20 MHz to 300 GHz. So we can measure radio radiation from astronomical objects on the Earth's surface only in these ranges. Radiation with a

Fig. 3.9 Course of the temperature in the earth's atmosphere

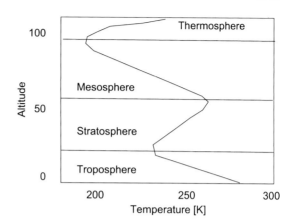

frequency below 20 MHz is reflected. This is important for the propagation of radio waves on the Earth's surface. Since these reflection properties depend on the state of the ionosphere, the propagation of radio waves can be sensitively disturbed by strong solar eruptions.

▸ In which frequency ranges can you communicate with the ISS space station and why?

Global Circulation System and Climate The weather system on earth is determined by the global circulation system. Roughly explained, it works like this: At the earth's equator the air is heated and rises upwards, near the ground colder air flows to the equator. Since the earth rotates, the *Coriolis force* acts on the moving air masses. *This can be* can be easily explained. Assume that on a rotating platform there is a small cannon at location 0 that is supposed to shoot a projectile. If the platform does not rotate, the bullet lands at A. If the platform does rotate, however, it lands at B, one has the impression that a force F_c has influenced the path of the bullet. This illusory force is called the Coriolis force and it deflects the winds to the right in the northern hemisphere of the Earth and to the left in the southern hemisphere. This is shown in Fig. 3.10.

The term *climate* covers effects of the earth's atmosphere (temperature, pressure, humidity, wind speed, etc.) that change over a period of several decades. Climate changes have always existed on earth. According to the theory of Milankovic (1879–1958), a change in the inclination of the Earth's axis (Fig. 3.11) of a few degrees (at present it is inclined 23.5° from the perpendicular to the Earth's orbital plane) and a small change in the distance Earth-Sun caused the ice ages of the past millions of years. Both variations are caused by the influence of the planets as disturbing forces. The greater the inclination of the Earth's axis, the more pronounced the differences between the seasons.

Seasons The seasons are caused by the different heights of the sun in summer and winter. In summer the sun is high in the sky and the days are long, in winter it is low and the days are short. At the time of the equinoxes, the Earth is illuminated from pole to pole

Fig. 3.10 In a rotating system, the Coriolis force causes the bullet fired at 0 to land not at (**a**), but at (**b**)

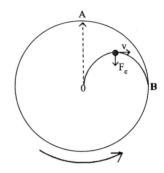

Fig. 3.11 The obliquity of the Earth's axis changes with a period of 41,000 years. (NASA)

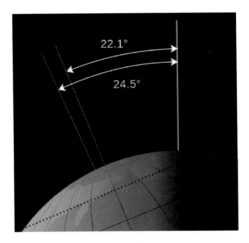

(Fig. 3.12). This is the case at the beginning of spring or autumn. At the beginning of winter (winter solstice) the sun is lowest and the days are shortest, at the beginning of summer it is highest and the days are longest.

So the seasons on earth are not caused by the different distance earth-sun due to the ellipticity of the earth's orbit. Currently, we are almost four million km closer to the Sun in early January than in early July. This has virtually no effect on the intensity of solar radiation, but we do notice something about it: for the northern hemisphere of the Earth, the summer half-year is slightly longer than the winter half-year. Near the Sun, the Earth moves faster around the Sun due to the Sun's greater gravitational pull.

Digression How can this be explained? Let us recall the first chapter, where we saw that the orbit of a planet around the Sun is stable when the centrifugal acceleration due to its orbit is equal to the gravitational acceleration due to the Sun:

Fig. 3.12 Earth at the times of
the equinox, day and night.
(NASA)

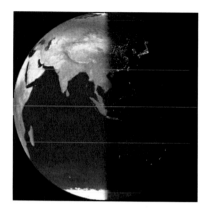

$$\frac{v^2}{r} = G\frac{M_\odot}{r^2}.$$ (3.3)

Here v is the speed of the planet on its orbit, r is the distance from the planet to the sun, $M_\odot$ is the mass of the sun. If, for example, the Earth is closer to the Sun at the beginning of January (the point closest to the Sun is called perihelion), then r is smaller.

- Centrifugal acceleration $\sim 1/r$,
- Acceleration due to solar attraction $\sim 1/r^2$,
- one can delete the r from the formula 3.3 $\rightarrow$ acceleration from the sun becomes larger, planet must move therefore faster around the sun.

Therefore, for the northern hemisphere of the Earth, winters are shorter than summers.

The seasons are caused by the inclination of the earth's axis to the normal. In summer, the sun is high above the horizon for us, the days are long, the earth heats up for a long time. However, the warming reacts hesitantly. Actually, it should be hottest around June 21, when the sun reaches its highest point in its orbit. But at this time, the earth and especially the oceans have not warmed up enough. That's why the hottest days here in the northern hemisphere are not until July or early August.

The Earth's atmosphere has changed in composition over time. It was only after life evolved on Earth about 3.5 billion years ago (presumably deep in the water to be protected from UV radiation) that oxygen was released through photosynthesis and slowly the Earth's atmosphere became enriched with free oxygen, eventually leading to the formation of the ozone layer. The evolution of life then continued explosively about 500 million years ago (Cambrian explosion). During the last 500 million years, there have been several episodes of mass extinctions. Within a short period of time, fossils indicate that about five times up to 80% of all life on Earth was wiped out. The last time this happened was 65 million years ago, known in science as the KT event (K stands for Cretaceous, T for

Tertiary) or popularly known as the extinction of the dinosaurs. We will come back to this
in more detail.

▸ Earth is the largest of the terrestrial planets. The tilt of the Earth's axis creates the
seasons. The Earth's atmosphere protects against extreme temperature contrasts. The
magnetic field provides a shield from charged particles from the sun and is created by
currents in the Earth's liquid core. The continental plates are in constant motion (tectonics).

3.2.2   Mercury

Mercury is the closest planet to the sun and because of its proximity to the sun it is difficult
to observe from Earth (in mid latitudes the planet can be seen on average about twice a year
as a morning or evening star). There are many famous astronomers who have never seen
Mercury in their lives. Since Pluto no longer counts as a planet, Mercury's role as the
smallest planet in the solar system falls to it. Its orbit is highly elliptical and so its distance
from the Sun varies between 46 and 70 million km. From Earth, Mercury can be seen in the
telescope as a small disk, which shows phases like our moon, but surface details cannot be
made out. Thus, Mercury's nearly 59-day rotation period has only been determined by
radar measurements (radar signals were sent to Mercury and the reflected signals measured;
as a result of rotation, they appear Doppler-shifted). The first images of Mercury's lunar-
looking, crater-strewn surface were taken in 1974 (US probe Mariner 10). Mercury was
mapped in more detail by the Messenger (Mercury Surface Space Environment Geochem-
istry and Ranging) probe. Messeneger was launched in 2004, placed in orbit around
Mercury in 2011, and crashed on Mercury's surface in 2015 after running out of fuel;
leaving behind a 15 m crater. Mercury's makeup can be described this way: a relatively thin
crust and a large metallic core. Mercury would be the ideal source of metals. The radar
reflections found in 1992 on the surface of the planet, which can be interpreted as frozen
water, i.e. ice, caused a great stir. At the polar regions of Mercury there are – similar to our
moon – areas where never a sunbeam penetrates into the interior of the craters.

After the two NASA missions Mariner 10 (1974, 1975) and Messenger (2011–2015),
ESA (European Space Agency) launched the BepiColombo mission in 2018 (Fig. 3.13).
Named after the Italian explorer Guiseppe Colombo (1920–1984), who devoted himself
intensively to the study of Mercury, the mission will enter Mercury orbit after 7 years of
flight time. This long flight time is necessary because the probe will get a gravity assist
several times due to close flybys of Earth and Venus, i.e. it will be nudged by the
gravitational fields of these planets, so to speak, and thus its orbit will be redirected towards
Mercury's orbit without the need for fuel. It will eventually orbit Mercury and provide data
of its surface and magnetic field for at least a year. BepiClombo will be exposed to extreme
temperatures of up to 350 °C.

Mercury has another record in the solar system besides its proximity to the sun: It is the
planet with the largest temperature contrasts between day (700 K) and night (100 K). So it
cools down by 600° during Mercury's night!

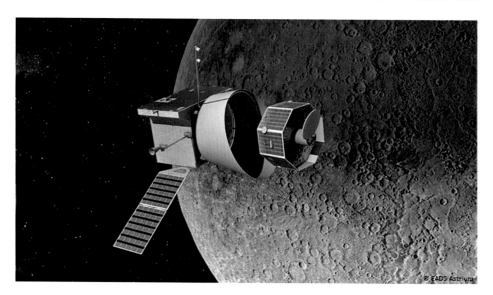

Fig. 3.13 The Mercury probe BepiClombo. (ESA)

Images of Mercury can be found in Figs. 3.14 and 3.15.

Life seems to be impossible on this planet. Why is it interesting at all to deal with Mercury so intensively? Mercury is an example of a planet that is exposed to extremely strong influences from the sun. Since our sun was more active in the early days of the earth than it is today, we can use the example of Mercury to understand how the solar wind and the short-wave radiation bombarded or changed the surface of the earth. In addition, many extrasolar planets have been found in the last two decades. Many of these planets are very close to their parent star, a large number even closer than Mercury is to the Sun. For this reason, Mercury is an excellent place to study these influences, which are essential for the formation of life.

▸ Mercury is the planet closest to the Sun with the largest temperature contrasts and it has a relatively large metallic core.

3.2.3   Venus

In terms of diameter, Venus could be called Earth's sister. No planet can come as close to the Earth as Venus (up to 42 million km). But all other data of Venus give at first a completely opposite picture of the Earth. Observations from the earth let recognize a large Venus disk, which shows in addition very beautiful phases, otherwise one sees no surface details. Because of the closer proximity to the Sun, the mean distance Venus-Sun is only 2/3 of the distance Earth-Sun, a dense cloudy atmosphere was assumed there (Fig. 3.16). Can there be life beneath the dense clouds? The results of the satellite probes were more

Fig. 3.14 Mercury. Image: Messenger probe. (NASA)

than sobering. Using radar, the surface of Venus was scanned and a surface covered with craters with some mountains (Maxwell Montes) could be seen. Particularly remarkable also the Venus rotation, it lasts 243 days, and the planet rotates *retrograde,* i.e. in the reverse sense to its movement around the sun.

While the Americans focused on the exploration of Mars in the sixties and seventies of the twentieth century, Russian spaceflight tried to explore Venus. Soviet spaceflight succeeded in landing several probes softly on the surface of Venus (Venera missions). A surface image can be seen in Fig. 3.17. One can recognize small stones. The environmental characteristics at the landing site of the probe are anything but inviting. The measured temperature is 740 K and the pressure at the surface of Venus is 90 times the pressure at the surface of the Earth. Such a pressure would correspond to the pressure in about 1000 m sea depth. The reason for this is in any case the dense Venus atmosphere. The high temperature can of course not be explained by the planet's closer proximity to the sun, but is a consequence of the extreme greenhouse effect. Venus' atmosphere consists largely of carbon dioxide CO_2.

Figure 3.18 shows a map of Venus produced by radar scans from NASA's Magellan satellite, and a size comparison of Venus-Earth.

The troposphere of the Venusian atmosphere reaches up to 50 km altitude. Between 30 and 60 km altitude one finds clouds of sulfuric acid droplets. Apart from these clouds, conditions at an altitude of about 53 km are pleasant: room temperature and a pressure of only 0.5 bar. Could primitive life have evolved? It is interesting to note that the atmosphere

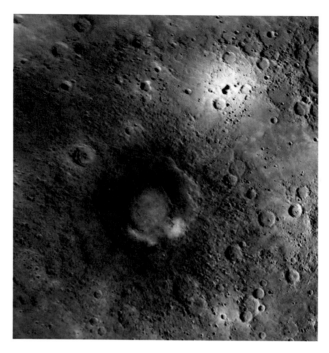

Fig. 3.15 Color-enhanced image of Mercury's surface (Messenger spacecraft, 2011). The colors can be used to infer the composition and evolution of the surface details. (NASA)

of Venus rotates around the body of Venus in only four days. The weather on Venus is almost identical at all locations and shows hardly any variations.

There has always been speculation about life in the atmosphere of Venus. As early as 1978, the US space probe Pioneer found particles whose size corresponded to terrestrial microbes when diving into the Venusian atmosphere. Later, signatures were also discovered in the spectrum of Venus. In 2020, the discovery of the gas monophosphane, also known as phosphine, was reported from observations with the radio telescope ALMA (an array of several radio telescopes up to 25 m in size in the Chilean Atacama Desert). Later, however, it became clear that the concentration of this gas was much lower than originally thought. In addition, phosphine could also be produced by abiotic but as yet unknown processes. Perhaps the flybys of the Mercury probe PepiColombo could provide better data. In any case, the question of whether there are low forms of life in Venus' atmosphere remains unanswered.

▶ Venus is similar in size to Earth, but due to its dense carbon dioxide-rich atmosphere it has a very high surface temperature, above the melting point of lead. Its rotation and surface characteristics could be determined by radar scans from satellites and from Earth. Venus was also visited by the US space probe Galileo (1989), which gained the necessary momentum to Jupiter by a close flyby of Venus, as well as by the Cassini-Huygens mission (1997), whose destination was the Saturn system.

Fig. 3.16 The surface of Venus, hidden by dense clouds, can only be scanned by radar measurements, even with space probes. (NASA)

Fig. 3.17 The surface of Venus, photographed on 5.5.1982 by the Soviet Venera 14 probe, which made a soft landing on Venus. (nssdcftp.gsfc.nasa.gov)

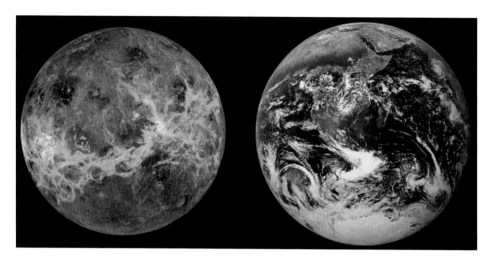

Fig. 3.18 Map of Venus obtained by radar scans using the US Magellan satellite. Earth next to it for size comparison. (NASA)

3.2.4   Mars

The red planet Mars has already attracted special attention in ancient cultures. No planet is subject to such strong changes in brightness as Mars. At a near-Earth opposition it can even surpass Jupiter in brightness for a short time. Near its conjunction with the Sun, it will be almost indistinguishable from other stars in the sky by an inexperienced observer. Especially when very bright, Mars shows a reddish hue, hence probably its association with the god of war (Ares to the Greeks, Mars to the Romans). From earth observations one recognized earth-similar surface structures, particularly strikingly the white polar caps, which change depending upon season.

Martian Canals?   In the early twentieth century, observers (mentioned as early as 1877 by Schiaparelli as canali) thought they could see a network of fine channels on its surface. This was interpreted as evidence of intelligent life on Mars. The Martians were thought to distribute scarce water supplies through an intelligent system of canals on the planet's surface (Fig. 3.19). Based on the migration of surface structures, it was also possible to determine the rotational period of Mars which is only about half an hour longer than that of Earth, and the inclination of the rotational axis of Mars is also similar to that of Earth, leading to the seasons mentioned above. All this contributed to the famous idea of the little green men on Mars.

In 1965, the US probe Mariner 4 was able to send the first images from near the surface of Mars to Earth. People were eagerly waiting to find out what the Martian channels were really all about. The images were all the more disappointing: The Martian surface

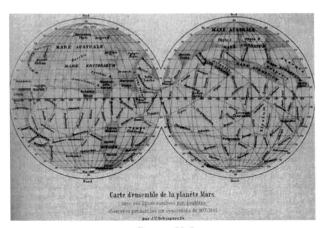

PLATE VII

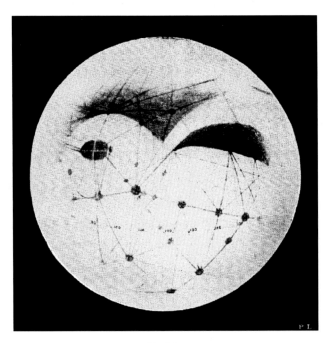

MARS
LONGITUDE 120° ON THE MERIDIAN

Fig. 3.19 Map of Mars based on observations by Schiaparelli, who thought he saw a canal system. (Wikimedia)

resembled a desolate desert littered with craters, and there was no trace of Martian canals. They were obviously an optical illusion, caused by the air turbulence in the Earth's atmosphere.

Fig. 3.20 Image of Mars taken with the Hubble Space Telescope HST. You can see the polar cap as well as cloud structures. (Hubble Space Telescope)

Probes Land on Mars: In the meantime, our image of Mars has changed again (Fig. 3.20). Probes have been successfully landed softly on its surface several times, and small or larger robotic cars have explored small parts of the Martian surface or conducted experiments to search for possible microbes. The atmosphere of Mars is very thin, and therefore its heat-retaining effect is less than that of Earth: temperatures around zero degrees Celsius are possible during the day, but at night temperatures can drop as low as -100 °C. Due to the low pressure of the Martian atmosphere of only about 1% of the pressure on the Earth's surface, no water can currently exist in liquid form on the Martian surface, it would evaporate immediately.

▸ Water boils at the earth's surface at about 100 °C, but below 100 degrees at higher altitudes because the atmospheric pressure decreases. For every 300 m of altitude, the boiling temperature decreases by about 1 degree. Boiling eggs on Mount Everest is therefore a difficult matter, it would boil at 70 °C and egg whites no longer curdle only the yolk.

However, structures have been found on Mars that are very reminiscent of dried-up river courses on Earth. In frozen form, water (in addition to frozen carbon dioxide) is also found at the poles and most likely below the surface. Therefore the dried up river courses could be witnesses of a climate change on Mars. With a much denser Martian atmosphere, water could well have flowed on the surface. In 1976 probes landed on Mars for the first time,

Fig. 3.21 Martian crater with ice in the center and on the walls. (Commons Wikimedia)

Viking 1 and Viking 2. The search for life was unsuccessful. In the year 2006 it was announced, one found a proof for short-time flowing water at a Mars crater. Images of a crater wall taken in 2001 were compared with images taken in 2005 (see also Figs. 3.21 and 3.22).

A very nice image of a Martian crater with ice inside and on its walls is shown in Fig. 3.21.

The "Martian face" (Fig. 3.23) caused great excitement. Some bizarre interpretations suspected that this was a kind of monument to Martian civilisations that had long since ceased to exist. A closer analysis of the image under different lighting conditions and with improved cameras revealed the Martian face to be a purely random arrangement of boulders. At the bottom right of the image, you can see an earlier image with a much poorer resolution. Here the impression of a Martian face is better.

Martian Meteorites On Earth, a meteorite was found in Antarctica that is clearly surface material from Mars. This is nothing unusual. When a meteorite strikes the surface of Mars, Martian soil material can be accelerated to such an extent that it reaches escape velocity and then crashes to Earth by chance thousands or millions of years later. However, it was believed to have found traces of Martian fossils in this Martian meteorite. This would be proof that there had been life on Mars at least once. However, this interpretation of the structures is doubted today; it is rather assumed that these are normal deposits (Fig. 3.24).

One more sight for future visitors to Mars from Earth must be mentioned. On Mars we find Olympus Mons, over 20 km high, the largest volcano in the solar system, which may have been active until about one million years ago (Fig. 3.25).

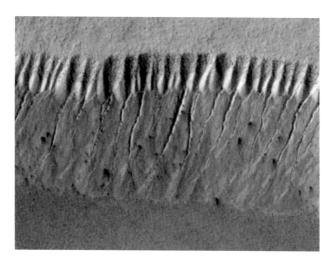

Fig. 3.22 So-called gullies at the rim of a Martian crater are seen as evidence for the intermittent presence of liquid water on Mars even today. Mars Global Surveyor Mission. (Commons Wikimedia)

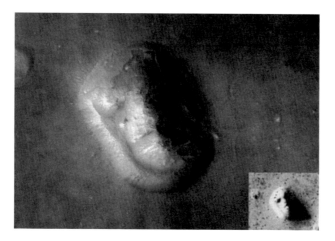

Fig. 3.23 The Martian face, a monument to a long-sunken Martian civilization? NASA

A surface panorama of the landing site of the Pathfinder mission is shown in Fig. 3.26. The rover conducted 15 chemical experiments and provided weather data over a period of 2 years. The Mars rover Curiosity landed on the surface of Mars in August 2012 and is the size of a small car. Unlike its predecessors, it is not powered by solar cells but by a radionuclide battery. This makes it independent of Martian weather. In addition to images, numerous investigations and test drillings are also taking place (Fig. 3.27).

▸ Mars is about half the size of Earth and the most similar to Earth of all the planets. There is ice on the polar caps as well as a thin atmosphere. The climate on Mars may have

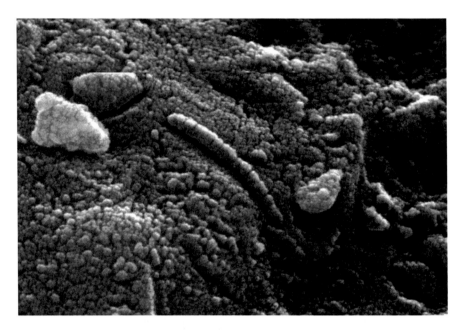

Fig. 3.24 The Martian meteorite found in Antarctica, said to contain traces of life. (NASA)

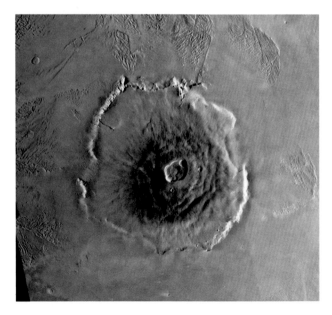

Fig. 3.25 The largest volcano in the solar system, called Olympus Mons, is located on Mars. (NASA)

Fig. 3.26 Surface of Mars, landing site of the Pathfinder mission. (NASA)

Fig. 3.27 Surface of Mars, panoramic image from the Curiosity rover. (NASA)

been warmer in earlier times, which seems to be evidenced by formations reminiscent of dried river valleys. The search for life forms on Mars has so far been unsuccessful.

A topographic map of Mars is shown in Fig. 3.28. The colors indicate different altitudes. Red means 3–8 km above mean Martian level, blue means up to 8 km below mean level. The first two Viking missions were in the blue areas because the surface is less densely littered with craters.

3.2.5   Future Mars Missions

The first flight of a mini-helicopter, Ingenuity, in April 2021, which hovered in the thin Martian atmosphere for about 40 s, caused a particular stir. However, a manned mission to Mars is unlikely in the near future, on the one hand for cost reasons, on the other hand also because of the high risk of such a mission.

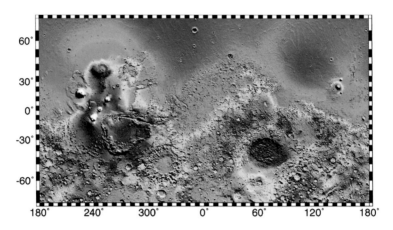

Fig. 3.28 Topographic map of Mars. (NASA)

3.2.6   Summary: Why So Different?

After all terrestrial planets have been discussed, we give a summary or a discussion why these planets are similar to each other on the one hand, but on the other hand show completely different properties:

Earth is currently the only planet with water in liquid form on its surface. Only Mars might have geyser-like eruptions of water on a small scale from time to time. During the formation of the Earth-like planets, Mercury was too close to the Sun, water could not persist there, and any very small deposits of water from Mercury's early phase would certainly have evaporated very quickly. However, one must remember that the early young Sun had only about 70% of its present luminosity. Venus may well have been covered by a global ocean of water in its early stages of development. What happened to this water? As already indicated, the Sun became more luminous as it evolved. Water began to evaporate, and the Sun's short-wave UV radiation split the water molecules into the lighter hydrogen and the heavier oxygen. Venus' gravity was not enough to hold the hydrogen, and the oxygen combined with rocks on the surface. This process is called photolysis of water. The dense atmosphere of Venus, with its high CO_2 content, also contributed to this process. On Earth, part of the carbon dioxide was removed from the atmosphere by tectonic processes.

Mars may also have had an ocean in early times. Computer simulations show that the rotation axis of Mars is subject to strong fluctuations, which has led to major climate changes. In any case, at present on Mars water in significant deposits in the liquid phase is not possible.

Fig. 3.29 Perseverance, the rover delivered to the surface of Mars in February 2021, is the largest Mars mission to date. (NASA)

If there has been water on Venus as well as on Mars, the question of the origin of life there naturally suggests itself. Life, at least as we know it from Earth, can only develop in the presence of water. Water serves as an important solvent, without which organic complex structures would be unthinkable. Was the time until the oceans on Venus and Mars disappeared sufficient for the development of at least primitive life forms? If so, could simple life have saved itself into some niche and survived the planetary changes? The suppositions are not quite so far-fetched, for we know that bacteria can survive the flight to the moon and back again without protection.

Another possibility, of course, would be that our Earth has long since infected these planets with life, even if only in very primitive forms.

It is hoped that the Mars mission Perseverance, which in February 2021 successfully set down an SUV-sized Mars rover softly on the surface of Mars (Fig. 3.29), will provide further information as to whether life exists on Mars. The main focus will also be on searching for signs of possible past life in a dried Martian lake in Jezereo Crater. A Mars helicopter will be used for the first time. This will be used to make small flights over the surface of Mars during a 30-day window. Figure 3.30 shows the Mars rover approaching the surface of Mars. The rover is braked by parachutes, which are connected to the rover by the Skycrane. There is a camera in the skycrane that took this picture.

Fig. 3.30 Mars rover Perseverance in the approach phase, shortly before landing. Image taken from the Skycrane. (NASA)

3.3   The Giant Planets

This section is about the giant planets, which are much larger than the Earth-like planets and have no solid surfaces.

3.3.1   Jupiter

Jupiter is the largest planet in the solar system. Its mass is about 300 Earth masses or 1/1000 of the mass of the Sun. Its diameter is 10 times that of the Earth. Its chemical composition is similar to that of the Sun, being 98% hydrogen and helium. With a telescope you can see from the earth a clear flattening of the planet as well as dark bands lying parallel to the equator and brighter zones. Furthermore, in the southern hemisphere you can see a huge area of high pressure, the great red spot, whose extent is 25.000×12.000 km.

In 1989, the Galileo spacecraft was launched, reaching Jupiter on December 7, 1995. The flight to Jupiter was not a direct one, but took place through several deflections during close flybys of Venus and Earth. This gave the probe additional acceleration toward Jupiter. Such flight maneuvers are called "gravity assist." The space probe penetrated Jupiter's atmosphere and measured its temperature and density.

Fig. 3.31 Details of Jupiter's atmosphere with large red spot. Image taken by Voyager probes. (NASA)

Jupiter's atmosphere (Fig. 3.31) consists of a stratosphere above the level $h = 0$ km, which extends to infinity, below which is the troposphere. In the troposphere there are clouds of different composition. Ammonia ice clouds at altitudes between 40 and 20 km, water ice clouds at altitudes between -70 and -60 km, droplets of water and ammonia between -90 and -80 km. Below -100 km gaseous hydrogen, helium, methane, ammonia are found. The clouds are very strongly coloured, water clouds and ammonia clouds are white, impurities (from sulphur, phosphorus and other compounds) determine the colouring. Various organic compounds are also found. The chemistry of the atmospheres of Jupiter and Saturn is very complex.

Digression Different parts of a planet's atmosphere can be studied using the radio occultation method. For example, a satellite measures the radio signals that pass through different parts of the atmosphere during the orbit of another satellite around the planet. As it passes through the different layers of the atmosphere, there is different absorption depending on the composition in each layer. Incidentally, this method is also used to closely study or monitor the state of the Earth's high atmosphere.

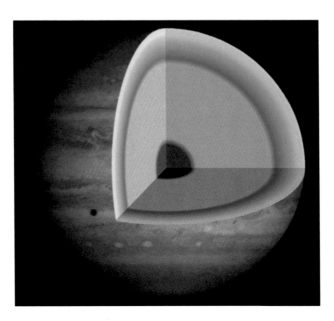

Fig. 3.32 The internal structure of Jupiter. At the center is the core, about the size of the Earth, then the regions of liquid metallic hydrogen and helium, then the zone of liquid hydrogen and helium. (Wikipedia)

On Earth, the atmospheric pressure at sea level is about 1 bar. On Jupiter there is a zone in the atmosphere, which becomes increasingly dense towards the inside, where this pressure prevails. Further inwards the hydrogen becomes metallic. The atoms lose their electrons and these are free to move, so currents can flow as in a metal. The core of Jupiter (Fig. 3.32) is made up of rocky material and is likely to be about the size of the Earth or a little more. The temperatures of these core regions are likely to be around 20,000 K. The transition from neutral to metallic hydrogen occurs at about 10,000 K and at a depth of about 20,000 km. The flow of currents in these zones explains the origin of Jupiter's magnetic field (Fig. 3.33). The strength of Jupiter's magnetic field is 20,000 times the strength of Earth's magnetic field. It extends up to 100 Jupiter radii, and it is compressed like the Earth's magnetic field on the side facing the Sun.

Jupiter, like Saturn, also has a ring system, but it is not as distinct as Saturn's and was only discovered by space probes.

Jupiter generates more heat than it receives from the sun. The reason for this is thought to be the cooling of its core.

Jupiter takes about twelve years to orbit the sun. Since it moves along the ecliptic, along which the twelve zodiacal constellations are arranged, it is in a different constellation each year. In 2021, for example, Jupiter is in the constellation Capricorn. In subsequent years, it then moves to the constellations of Pisces, Aries, etc. Jupiter is the planet without seasons, its rotation axis has only a very slight inclination.

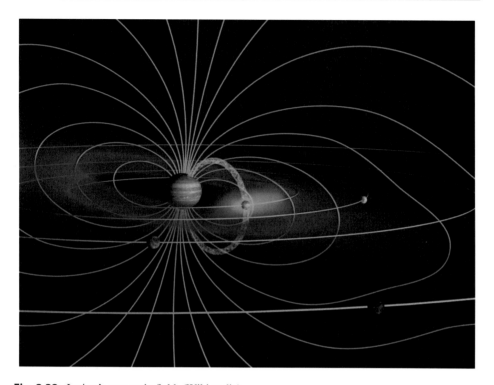

Fig. 3.33 Jupiter's magnetic field. (Wikimedia)

▶ Jupiter is about ten times the diameter of Earth and is the largest planet in the solar system.

Jupiter is expected to receive another visit from a probe in 2029: In 2023, ESA's JUICE mission (Jupiter Icy Moon Explorer) is scheduled to launch, primarily to study Jupiter's icy moons.

3.3.2 Saturn

Saturn is the faintest of all planets visible to the naked eye. Nevertheless, it reaches the brightness of the brightest stars. Even with a small telescope, Saturn's rings can be seen (Fig. 3.34). They consist of many meter- to centimeter-sized ice particles orbiting Saturn in the equatorial plane. Resonance effects with Saturn's moons cause gaps in the rings, named after astronomers. The *Cassini division* can be seen with smaller telescopes, further there is the Encke division and others. Images taken with space probes show that there are in fact a large number of rings. The extension of Saturn's rings is about 62,000 km, the thickness however only few 100 m. Since the orbits of Saturn and Earth are inclined to each other, from Earth one sees the rings once from "above" and once from "below". For short periods

Fig. 3.34 Saturn with rings, the Cassini division is clearly visible. In the southern hemisphere auroras are visible. (Source: Hubble Telescope, NASA)

of time it can also happen that we look exactly at the very thin edge of the rings, then Saturn appears in the telescope without rings, only a dark stripe is visible. This happens about every 15 years. In the year 2025 we look again at the edge of the ring, then we see Saturn without rings for a few weeks.

The formation of rings is explained as follows: At a certain distance from a planet, larger bodies are torn by tidal forces. One can easily calculate this zone, which is called the *Roche boundary,* by equating the tidal force acting on a body with the force of attraction with which two imaginary mass halves of the body attract each other. The rings of Saturn are within the Roche limit, as are the rings of Jupiter and the other two gas planets, Uranus, and Neptune. Larger moons that approach the Roche boundary are torn apart by tidal forces and become small particles of ice. Why are Saturn's rings seen so spectacularly in the telescope, while the rings of the other gas planets were not discovered until much later? Saturn's ring particles consist mainly of ice, which strongly reflects sunlight, while the other planetary rings consist of dark material (in the case of Jupiter, dust particles).

The Spitzer Space Telescope observes mainly in the infrared. In 2009, this instrument was used to discover yet another ring of Saturn, made of material from Saturn's moon Phoebe, which extends between 6 and 12 million km. Viewed from Earth, this ring would be twice the size of the full moon in the sky. Then in 2015, the WISE space telescope detected an even greater extent of this ring – it reaches up to 16 million km Saturn's distance. In Fig. 3.35, you can see an artist's rendering of the ring. Saturn itself is the small dot on the left, Saturn with its main rings is zoomed out on the right.

On Saturn, whose axis of rotation is inclined at about 27°, seasonal effects are observed, but the seasons last more than 7 years.

Fig. 3.35 Saturn's outermost ring, visible only in the infrared, made of material from the moon Phoebe. (Credit: NASA/JPL-Caltech/Keck)

Saturn's atmosphere is also very dynamic, with hot spots observed at the poles caused by gas reaching there, which is compressed and heats up as a result, as well as several thousand kilometre hurricane-like areas. Every 29 years the great white spot appears in the northern hemisphere.

In size, Saturn is slightly smaller than Jupiter, it is the most oblate of all the planets in the solar system. The equatorial diameter is a little more than 120,000 km, the pole diameter is about 107,000 km.

The rotation period is below ten hours. You can see fewer structures in Saturn's atmosphere than on Jupiter. Saturn needs about 30 years to orbit the sun. In the interior of Saturn, which is similar to the interior of Jupiter, heavier helium droplets sink to the bottom, thus gravitational energy is released. Saturn therefore generates more heat than it receives from the Sun. Another peculiarity of Saturn should be mentioned: of all the planets in the solar system, it has the lowest density of only $0.7 \mathrm{g/cm}^3$. Saturn would therefore be swimming in a huge bathtub filled with water.

▶ Saturn is especially known for its spectacular ring system, which is already easily visible with small telescopes. The rings consist of ice particles up to about a meter in size.

Fig. 3.36 Uranus with rings, photographed in infrared light, where the dark rings are visible. (Image: Hubble telescope)

3.3.3 Uranus and Neptune

New Planets Are Discovered Sometimes the two planets Uranus and Neptune are also called ice giants. Uranus was found in 1781 by W. Herschel. Under extremely good conditions you could still see this planet with the naked eye, binoculars will definitely do, but the planet was unknown in ancient times. The story of Neptune's discovery is even more exciting. From a detailed analysis of the motion of Uranus, it was found that its orbit must be disturbed by an unknown planet. It was even possible to calculate approximately the location of the still unknown planet. This was done by the celestial mechanic Le Verrier. Now a race to find the planet by observation began, and it was the astronomer Galle who discovered Neptune at the Berlin Observatory in 1846. Galle was very modest and he thought that the discovery of Neptune was actually due to Le Verrier's calculations and that he should be listed as the discoverer of Neptune. Neptune can already be found with a small telescope. Both planets, Uranus and Neptune, appear as a small disk with small telescopes, on which, however, no structures can be seen (Fig. 3.36).

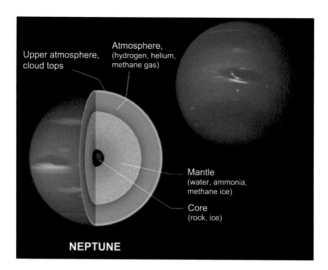

Upper atmosphere, cloud tops

Atmosphere, (hydrogen, helium, methane gas)

Mantle (water, ammonia, methane ice)

Core (rock, ice)

NEPTUNE

Fig. 3.37 The structure of Neptune and an image showing the blue planet. (NASA)

The masses of Uranus and Neptune are each about 15 Earth masses. Neptune's axis of rotation is inclined by 29°, but Uranus' is inclined by 98°. With an orbital period of more than 80 years, each pole of Uranus is therefore directed towards the sun for more than 40 years, the planet seems to roll around the sun, and on top of that Uranus rotates retrogradely, i.e. in the opposite sense to its motion around the sun. A peculiarity in the solar system. The rotation period of both planets is 17 h. Although Neptune is farther from the Sun than Uranus, it has about the same temperature, so it has an internal heat source. Therefore, Neptune's atmosphere appears more detailed (see also Fig. 3.37) than that of Uranus. Neptune is actually the blue planet in the solar system. Wind speeds of more than 2,100 km/h have been found in its atmosphere. Both planets also have a magnetic field.

With modern large telescopes and also from space one gets today very detailed pictures of the two planets. Both planets also have ring systems.

The Rings of Uranus The ring system of Uranus (Fig. 3.36) was only found by chance in 1977. Planets move in the sky. The further away, the slower this movement. During this migration, planets can also repeatedly occult stars. Similar to radio occultation methods, the composition of the atmosphere can be inferred by analyzing the starlight during stellar occultations, so such occultations are followed very closely. Shortly before a star was occulted by the Uranus disk (see also Fig. 3.38), several brief drops in brightness were detected, which were repeated after the actual occultation, i.e. they were symmetrical. This is simply explained by a ring system around Uranus. The rings of Uranus consist of very dark material. Because of the unusual inclination of the axis of rotation of Uranus, we see the rings around the planet, the planet appears in the middle.

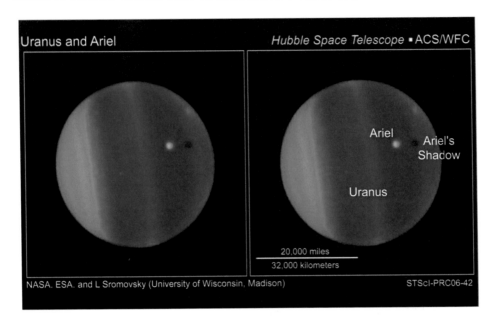

Fig. 3.38 Uranus, imaged with the Hubble Space Telescope. You can see the shadow cast by a moon on the surface of Uranus. There an observer would see a total solar eclipse

The ring system of Neptune is still little explored.

▶ Uranus and Neptune are also called ice planets, they are about half the size of Jupiter and Saturn, but are no longer visible to the naked eye.

3.3.4 Composition: Giant Planets

The four giant planets of the solar system are further subdivided into

- Gas planets: Jupiter and Saturn,
- Icy planets: Uranus and Neptune,

by virtue of their composition.

All of them have a ring system and many moons, which will be discussed in the next chapter. Except for Uranus, these planets also have an internal energy source, which can be explained by cooling or sinking processes in the planet's interior. These energy sources often cause very complex structures in the atmospheres.

Generally speaking:

▶ The exterior of the solar system is dominated by ice.

3.4   The Moons of the Planets

The two innermost planets, Mercury and Venus, have no moon. Our Earth has one very large moon compared to its diameter, Mars has two tiny ones, and the giant planets Jupiter, Saturn, Uranus, and Neptune have many moons, but they are tiny compared to their diameters. We will first discuss our moon and then the satellite systems of the other planets.

3.4.1   Our Moon

The Moon as a Celestial Body   The Moon is the closest celestial body to us and the only one that has been walked on by humans, the first time on 21 July 1969 (N. Armstrong). A total of 12 astronauts have walked on the moon (Fig. 3.39). Again and again one reads in various sources that the American moon landings were faked and only productions in studios. One of the arguments for these claims is that from Earth the left behind lunar landing vehicles cannot be seen in large telescopes. An easily comprehensible calculation shows that this is not possible even with the largest telescopes (currently ten meters in

Fig. 3.39 Apollo 11: On 21 July 1969, humans set foot on the Moon for the first time. (NASA)

diameter). The resolution of the telescopes is not sufficient for this, moreover the observations are strongly affected by turbulences in the earth atmosphere. Different satellite-missions, where also satellites were brought into lunar orbit, however did show quite clear the left-behind landing-vehicles.

Let's stay briefly with observations of the Moon from Earth. Already with binoculars one can see very large lunar craters (diameter larger than 100 km) and telescopes show a variety of craters and mountains, with good resolution also grooves and cracks on the lunar surface. However, the best time to observe the Moon with a telescope is not – as many think – around the time of the full Moon, but a few days before or after the first or last quarter. During this time the moon is about half illuminated and you can see impressively the craters and mountain ranges which cast shadows. When the moon is full, on the other hand, you see a largely featureless bright lunar disk. The moon is by far the most interesting observation object for small telescopes. During a near-Earth Mars position, about as many details can be seen on this planet as can be seen with the naked eye on the lunar surface (Fig. 3.40).

From Earth, we always see the same half of the lunar surface, since the Moon's rotation period is equal to its orbital period. The far side of the Moon has only been photographed by satellites in lunar orbit (first in 1959 by the Soviet probe Lunik 3). The Chinese probe Chang'e 4 successfully landed on the far side of the Moon in 2019 (Fig. 3.41). A rover was also deployed (Yutu).

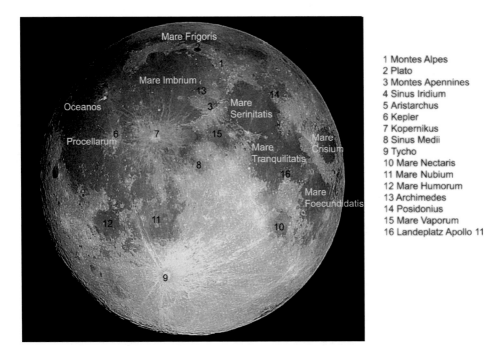

Fig. 3.40 Lunar map, the large lunar seas can be seen with the naked eye from Earth. (Photo: A. Hanslmeier, Private Observatory)

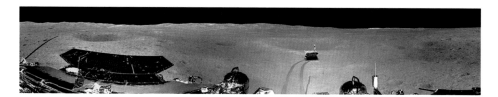

Fig. 3.41 Panormap image of Chang'e 4 lander with Yutu robotic rover. (Chinese. Space Agency)

Due to its low mass (1/81 of the Earth's mass), the Moon cannot hold an atmosphere. If gases escape from its surface, they immediately escape into space. From the earth you can see dark spots with the naked eye, which are called lunar seas (lat. maria). There are very imaginative names like the sea of rain (Mare Imbrium), the sea of honey (Mare Nectaris) and so on. There was also the idea that the moon was inhabited like the sun. Jules Verne (1828–1915) still has this idea. He describes the flight to the moon, where the space capsule is to be shot with a huge cannon. Today we know: The moon is dry as dust, there is no water. However, there seems to be ice at the polar regions. Sunlight never penetrates deep craters there and ice could remain there for several billion years. Such ice deposits would be very interesting for future space stations on the moon as a water supplier.

The composition of the rocks of the lunar surface is similar to that of the Earth's crust. Our moon was formed in the early phase of the solar system by the collision of the early Earth with a planet about the size of Mars more than four billion years ago. Lunar rocks are composed mainly of silicates and contain almost no metals or water. The only weathering that occurs to the lunar surface is due to

- extreme temperature contrasts between day and night,
- Irradiation of the surface by solar wind particles

Therefore, the footprints of the astronauts will be recognized for several 100,000 years.

On the lunar surface one finds different formations with telescopes: The aforementioned lunar seas consist of dark basalt lava that flowed from the impact of larger asteroids, the highlands (terrae) consist of feldspars and are littered with numerous crater impacts. The lunar seas are younger than the highlands. The lunar mountains were not formed by folding between plates as on Earth (Earth: e.g. the Alps were formed by the collision of the African plate with the European plate), but as a result of the impact of large asteroids. Lunar mountains have been named after terrestrial ones, e.g. the Alps or the Caucasus can be found on the Moon. Lunar craters are usually named after astronomers.

Seismometers left on the Moon do not indicate lunar earthquakes. The moon is geologically inactive, it has no iron core and no magnetic field. The moon always shows the same side to the earth, because its rotation duration is equal to its earth revolution duration. Therefore, the back side of the moon can only be explored by space probes. On the back side there are almost no lunar maria. This asymmetry can be explained easily: In the lunar maria, the density is greater, basalt is heavier than feldspar, and therefore this half of the Moon, where these areas are found, faces the Earth.

▶ Our moon is relatively large compared to the earth and was formed by a collision in the early days of the earth with another planet. It is the only celestial body (apart from Earth) that has been walked on by humans. The dark spots visible to the naked eye are huge basalt basins formed by large impacts.

- Moon: Diameter: 3476 km,
- Lunar orbit: major semimajor axis 384,400 km; near Earth 363,300 km (periapsis, perigee), far Earth 405,500 km (apoapsis, apogee); orbital inclination: 5.15°.
- Lunar surface: 38 million km^2, roughly equivalent to the area of Africa and Europe combined.

3.4.2 The Moons of Mars

Mars has two tiny moons, called Deimos and Phobos. Both are irregularly shaped celestial bodies that move very rapidly around the planet at a short distance from the surface of Mars. Phobos is only 2.8 Martian radii from the surface of Mars and moves around Mars in only 7 h 39 min (Fig. 3.42). Thus, it rises and sets several times in the course of a Martian day lasting more than 24 h. Deimos is 7 Mars radii away and takes about 30 h to complete one orbit. Phobos is about $27 \times 21 \times 19$ km across, Deimos slightly smaller, $15 \times 12 \times 11$ km. Phobos' orbit is unstable and it will crash into the surface of Mars.

These moons could be captured asteroids.

3.4.3 The Moons of Jupiter

Already Galileo Galilei recognized around 1609, when he first pointed a telescope into the sky to Jupiter, that this planet seems to be surrounded by four little stars that move around it. Galileo immediately interpreted these four little stars as moons of the planet Jupiter, and even today these moons, Io, Europa, Ganymede, and Callisto, are called the Galilean moons of Jupiter. These are particularly interesting, also for astrobiology, that is, the search for life in the universe.

Figure 3.43 shows a size comparison of these four moons with Jupiter.

Io Io is the innermost of the four Galilean moons. There are active volcanoes on Io (Fig. 3.44), ejecting sulfur or sulfur dioxide. There are also snowfalls of sulfur dioxide. A hot spot has also been found on Io, extending 200 km, where the temperature is 300 K. The cause of volcanism is heating due to tidal action. Io is at the same distance from Jupiter as the Earth's moon is from the Earth, but the mass of Jupiter is 300 times the mass of the Earth, and therefore Io is compressed and pulled apart. The tidal range is about 10 times that of Earth. Therefore, this moon is constantly being compressed and pulled apart, which causes heating as a result of friction and explains volcanism. The average orbital semi-axis is 421,800 km and Io moves around Jupiter in only 1.77 days. Why this fast movement?

Fig. 3.42 The Mars moon Phobos photographed in 2008 with the Mars Reconnaissance Orbiter. The 12 km large crater Stickney is visible on the *left*. (Credit: NASA/JPL)

Fig. 3.43 Size comparison of the four Galilean moons (Io, *left*), Europa, Ganymede, and Callisto *(right)* with Jupiter, the large red spot on Jupiter is clearly visible, a vortex larger than Earth that has been visible for more than 200 years. (NASA)

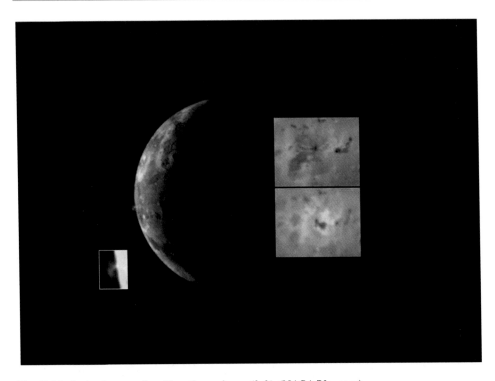

Fig. 3.44 Jupiter's moon Io with active volcano *(left)*. (NASA/Voyager)

Again briefly as a reminder: Due to its more than 300 times Earth's mass, Jupiter exerts a strong gravitational pull on Io, therefore the orbital velocity of this moon must be very high.

Io moves within the magnetic field of Jupiter and atoms of the higher ionosphere are ionized and therefore, as electrically charged particles, are subject to the forces caused by the magnetic fields which cause there to be a plasma tube of ions from Io around Jupiter.

Europe Europa is slightly smaller than our Earth's moon (cf. Fig. 3.45).

Already Earth-based observations and analysis of the reflected light from this moon of Jupiter indicated that its surface must consist essentially of water ice. This was then confirmed by space probes flying close by (Figs. 3.46 and 3.47). The surface temperature at Europa is $-150\,°C$. However, there is a very strong argument that beneath this ice shell is an ocean of liquid water with various salts dissolved in it. The ice sheet could be up to 100 km thick. Electric currents can be induced by the saline water and thus explain the measured magnetic fields or their variations. Why can Europe hold liquid water beneath its extremely cold surface? Again it is the strong tidal forces of Jupiter. This is also called tidal heating. The inner structure of Europa is sketched in Fig. 3.48.

Fig. 3.45 Comparison of the size of the Europa-Earth-Moon. (NASA)

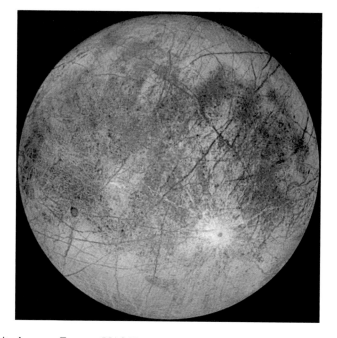

Fig. 3.46 Jupiter's moon Europa. (NASA)

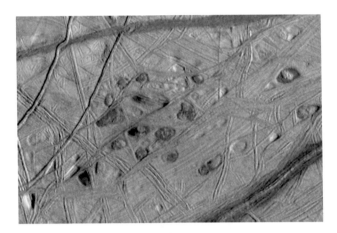

Fig. 3.47 Details of the surface of Europa. (NASA)

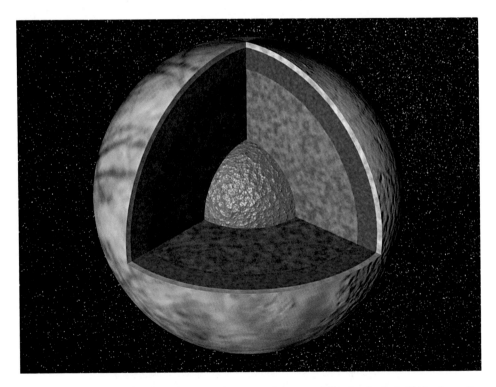

Fig. 3.48 The internal structure of Europe. An ocean of water and dissolved salts is hidden beneath a crust of ice. (NASA)

Europe's ocean beneath the ice crust shows many parallels to the conditions on the ocean floors of the Earth. There are the "black smokers", where mineral-rich hot water rises in geyser-like fountains, a consequence of volcanic activity. This is also where life is thought to have originated. Possible life in the Europa Ocean would also be protected from short-wave radiation by the thick ice crust. This is why Europa is one of the most important candidates in the search for life in the solar system. Organic compounds are also found on the surface. There were even plans to send a landing probe there, which would then drill through the ice crust to search for life forms in situ. This project has since been abandoned for cost reasons as well as fear of destroying life there. It is therefore hoped to find possible life forms indirectly (Fig. 3.49).

Ganymede and Callisto Ganymede is the largest moon in the solar system, like some other moons it exceeds Mercury in diameter. It has a magnetic field, which is also explained by a liquid ocean with salts below the surface. Callisto is completely differentiated, i.e. while it was still liquid in its formation phase, the heavier components sank downwards. Thus its core consists of sulphur and iron, followed by a silicate mantle, and finally the ocean and ice. Further, a very thin atmosphere consisting mainly of oxygen has been found. Callisto is also larger than our moon.

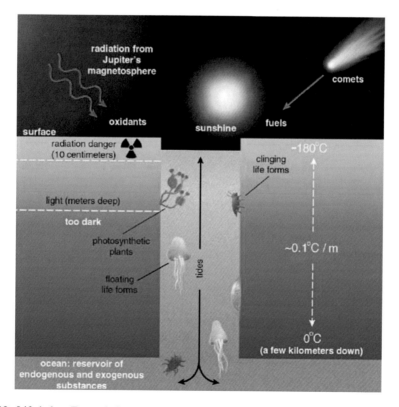

Fig. 3.49 Life below Europa's ice crust would be theoretically possible. (NASA)

Table 3.3 The four Galilean moons of Jupiter and data for our moon for comparison

Name	diam.	Mass	Density	Reflection
	[km]	Earth Moon = 1	[g/cm^3]	[%]
Callisto	4820	1.5	1.8	20
Ganymede	5270	2.0	1.9	40
Europe	3130	0.7	3.0	70
Io	3640	1.2	3.5	60
Earth moon	3476	1.0	3.3	12

A summary of the most important properties of the Galilean moons is given in Table 3.3.

3.4.4 Other Moons of Jupiter

A total of 79 moons of Jupiter are known today. The four Galilean moons are either larger than Mercury (Ganymede and Callisto) or about the size of our moon (Europa and Io), but small compared to Jupiter. The remaining moons are even smaller. The system of satellites of Jupiter is divided into:

- Galilean moons.
- Four moons of the middle group: the largest are Himalia (170 km) and Elare (80 km), distance 155–164 Jupiter radii, orbital inclinations up to 29°, eccentricity 0.13–0.21.
- Outermost group: 10–30 km in diameter. 290–332 Jupiter radii away, very large orbital inclinations (147–163°), and high eccentricity. This high orbital inclination and the partly retrograde orbital motion indicate that they are probably captured minor planets.

The four Galilean moons of Jupiter are roughly comparable to our moon in size, but otherwise completely different: Io has active volcanism, under the ice crust of Europa one suspects an ocean of liquid salty water; theoretically there could be life there.

3.4.5 Saturn's Moons

As of 2021, 82 of Saturn's moons were known. There are also some remarkable features in this system.

Titan Titan was already discovered in 1655 by Chr. Huygens. It was the sixth known moon of a planet at that time (besides Titan the four moons of Jupiter found by Galileo and our moon). It moves once around Saturn in a little more than 15 days and can easily be found with smaller telescopes as a star in the equatorial plane of Saturn (just extend the ring

Fig. 3.50 Artist's impression of the Cassini spacecraft with Huygens sample and Saturn in the background and Titan. (NASA/Cassini)

plane). The average density is only 1.88 g/cm^3, which also suggests water as an important component (in the form of ice and a possible ocean below the surface). Titan is the only moon in the solar system with a dense atmosphere. Even from the first space probes (Voyager), Saturn visited, one got no pictures of the Titan surface, because this remains hidden under dense clouds.

The US space probe Voyager 1 was launched in 1977 and reached the Saturn system in 1980. In February 2021 it was already at a distance of 152 AU from the Sun (a radio signal to Earth took about 20 h!). Radio communication with Voyager 1 is expected to continue until 2025. The probe is moving away from us by about 3.6 AU or 540 million km per year. But back to Titan.

The temperature on Titan's surface is $-180°$, and the pressure is about 1.5 times that on Earth's surface. As a result of Titan's lower gravity, we could fly there with our arms spread out. The atmosphere is mostly nitrogen and organic compounds make it so dense. Because of the pressure and temperature, it was thought that oceans of methane could be found on Titan's surface.

Testing all these assumptions was the task of the Cassini mission (Fig. 3.50). The launch took place in 1997. In order to get the necessary thrust for a quick trip to Saturn, the probe flew first past Venus and then Earth (1999).

Fig. 3.51 An image of Titan composed of three different colours, where surface structures can also be seen. (NASA/Cassini)

After seven years of flight, the probe reached the Saturn system in June 2004. Then, in January 2005, the European Huygens probe separated to land successfully on Titan's surface on 14 January. Due to a technical error, we only got a few images, which showed a surface at the landing site probably covered with ice lumps contaminated by dust (Figs. 3.51 and 3.52). Radar scans of Titan's surface revealed no oceans, but at least large lakes of methane (Fig. 3.53). This makes Titan the only celestial body in the Solar System besides Earth where a liquid is found on the surface (Figs. 3.51 and 3.52).

Thus, for the reasons mentioned above, Titan is a very interesting candidate, along with Mars and Europa, in the search for life in the solar system (Fig. 3.53).

Enceladus: Ice Volcanoes   Another surprise of the Cassini mission was the discovery of ice volcanoes on the surface of Saturn's moon Enceladus. In the case of cryovolcanoes, frozen substances such as methane, carbon dioxide, ammonia and water are melted in the interior of a moon by heating (e.g. tidal forces) and penetrate to the surface. Just below the surface of Enceladus there is liquid water, which rises in huge fountains (due to the low gravitational pull of the small moon Enceladus) up to a height of 500 km (Figs. 3.54 and 3.55). This would make Saturn's moon, which is only about 500 km across, another candidate in the search for life.

Fig. 3.52 Surface details at the
Huygens probe landing site.
(NASA/Cassini-Huygens)

Fig. 3.53 Methane lakes on Titan as determined by radar scans. (NASA/Cassini)

Fig. 3.54 The surface of Saturn's moon Enceladus, mostly covered with ice. (NASA/Cassini)

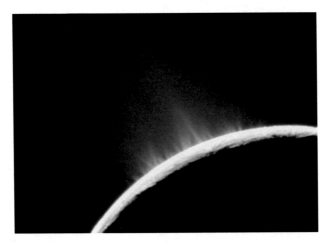

Fig. 3.55 Water fountains shoot up to 500 km from the surface of Saturn's moon Enceladus. (NASA/Cassini)

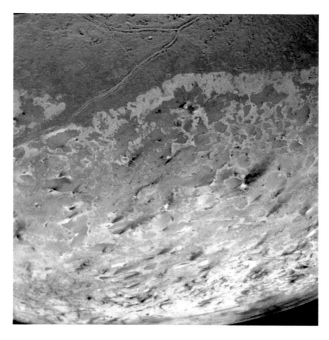

Fig. 3.56 Neptune's largest moon, Triton; traces of cryovolcanism can be seen. (NASA)

Japetus and Rhea These two moons of Saturn have diameters of about 1500 km each. The surface of Japetus has a dark and bright hemisphere. The dark half probably comes from dust kicked up during an impact on Saturn's moon Phoebe.

The other moons of Saturn are very small and partly only little explored and some are certainly – as with Jupiter – captured asteroids.

3.4.6 The Moons of Uranus and Neptune

Uranus has 27 moons, Neptune 14. Neptune's moon Triton has a diameter of 2700 km and it moves retrograde around Neptune. Possibly it was captured by Neptune. Its surface temperature is only 38 K and frozen nitrogen is found, as well as an ice crust and active cryovolcanism (Fig. 3.56). At present, Triton is only about 350,000 km from the surface of Neptune. It will still approach Neptune and, once it gets within the Roche boundary, will be torn apart by its tidal forces, which will happen in about 100 million years. Then a magnificent ring will form around Neptune.

3.4.7   Summary

The moons of the planets in the solar system are very different, and their exploration with space probes showed completely unexpected results. Lakes of methane on the surface of Titan, oceans of liquid water beneath the ice crusts of Europa, Ganymede and possibly deeper at Titan. Organic compounds in the atmosphere of Titan and on the surface of Europa, ice volcanism on Enceladus. In this way, some of the moons become interesting for the search for life.

Dwarf Planets and Small Bodies

4

In this section we first discuss the newly defined class of dwarf planets, then the minor planets or asteroids, the comets, and the matter between the planets, the interplanetary matter.

After reading the chapter, you will know what

- is the difference between minor planets, dwarf planets,
- why there are asteroid belts,
- whether the Earth is in danger from asteroid impacts,
- what comets are all about,
- which is why research into small solar system bodies (SSSBs) could secure our future needs for metals and other raw materials.

4.1 Asteroid Belt in the Solar System

The small planets or asteroids are not evenly distributed in the solar system, but are mainly concentrated in so-called belts.

4.1.1 Asteroid Belt, Main Belt

In older books, the term minor planetary belt always refers to the area between Mars and Jupiter, where most minor planets were first found. On New Year's Eve 1800, the astronomer Piazzi discovered the object Ceres between Mars and Jupiter. That there were objects there had been suspected for some time, since a gap appears between the orbits of Mars and Jupiter.

Digression One can represent the distances of the planets from the sun by a simple relation (Titius-Bode series):

$$a = 0.4 + 0.3 \times 2^n. \tag{4.1}$$

If one inserts $n = -\infty, 0, 1, 2, 3, \ldots$ one finds approximately the distances of the planets from the sun. For the index $n = 3$ there is no planet, but in this distance range is the main belt of minor planets.

How many objects there are in the main belt is unknown, but certainly several 100,000. The journey in a spaceship from Mars to Jupiter would be relatively safe, however, despite these many asteroids, because the probability of collisions with asteroids is extremely low, due to the large spatial extent of the belt. A sketch of the asteroid belt between Mars and Jupiter can be found in Fig. 4.1.

4.1.2   Kuiper Belt

Outside the orbit of Neptune one finds the objects of the Kuiper belt. This belt of objects was first suspected by Kuiper (1905–1973) in 1951, and the first object discovered in it was QB1 in 1992. This has a diameter of only about 200 km. Because of the greater distance of the Kuiper belt objects, fewer are known than from the main belt. A total of about 100,000 objects with diameters greater than 100 km are thought to exist between 30 and 50 AU solar distance.

The Kuiper belt objects (Fig. 4.2), like the main belt objects, are highly concentrated with respect to the ecliptic plane. Kuiper belt objects include numerous dwarf planets, the best known of which is Pluto, which along with other objects is listed as a major Kuiper belt object. The group of *centaurs* includes about 300 objects that migrated to more interior regions of the solar system due to the influence of Neptune's gravity. Some cross the orbits of Neptune, Saturn or Jupiter, so will collide with these planets within the next million years.

4.1.3   The Oort Cloud

If one analyzes the orbits of comets, one finds that they seem to come from all directions. Oort (1900–1992) suspected that comets might come from a cloud surrounding the solar system, which is now called Oort's cloud. There could be several billion comets in this cloud. An impression of the dimensions of this comet cloud is given in Fig. 4.3. In the

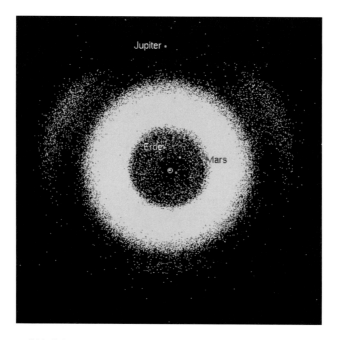

Fig. 4.1 The asteroid belt between Mars and Jupiter. (Creative commons; Rivi; cc-by-sa 3.0)

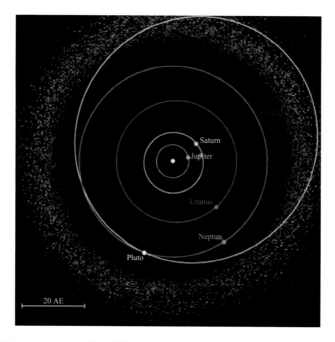

Fig. 4.2 The Kuiper belt and orbits of Neptune and the dwarf planet Pluto. (Source: MPIA)

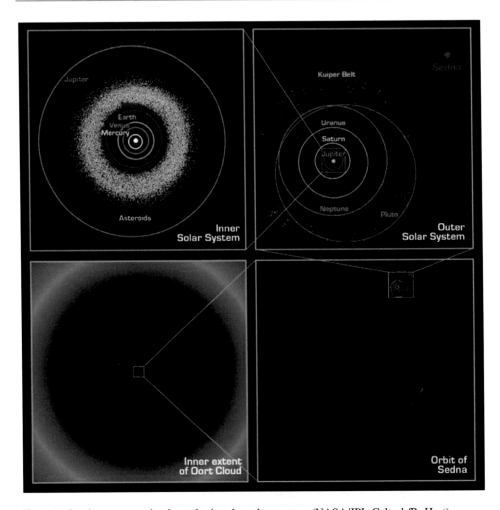

Fig. 4.3 Oort's cometary cloud enveloping the solar system. (NASA/JPL-Caltech/R. Hurt)

upper left corner the inner solar system up to the orbit of Jupiter is drawn with the main belt of asteroids between the orbits of Mars and Jupiter, in the upper right corner the Kuiper belt is shown, in the lower left corner the Oort cloud and in the lower right corner the orbit of the object Sedna.

▶ There are three belts of minor planets in the solar system: Main Belt, Kuiper Belt, and Oort's Cloud.

4.2 Minor Planets, Asteroids

4.2.1 Observation

The first minor planet discovered, Ceres, is now listed as a dwarf planet.

The exploration of the minor planets was initially only possible through brightness measurements from Earth. One notes a periodic brightness change, which arises from the rotation of the object. Minor planets are usually irregularly shaped, or their surfaces are not uniform. They therefore appear somewhat brighter when they show us their larger cross-sectional area as a result of rotation. From the color we can also get information about the rough surface texture.

Meanwhile, there have been numerous missions to minor planets. To our surprise, it turned out that there are relatively many minor planets with companions, moons. A very well-known example is the minor planet Ida with its tiny moon Dactyl (Fig. 4.4). It was imaged in 1993 by the Galileo spacecraft from a distance of about 10,500 km. The extent of the irregularly shaped object is about 53.6 × 24.0 × 15.2 km. Ida was discovered in 1884 by an astronomer of the Vienna Observatory (J. Palisa, 1848–1925). Its rotation period is 4.6 h, and the major semi-axis of its orbital ellipse is 2.86 AU. The trajectory of the Galileo spacecraft is drawn in Fig. 4.5.

There are also asteroids that come relatively close to Earth. On September 29, 2004, the asteroid Toutatis, about 4.5 × 2.4 × 1.9 km in size, passed Earth only about four times the

Fig. 4.4 Minor planet Ida with moon Dactyl (right). (NASA)

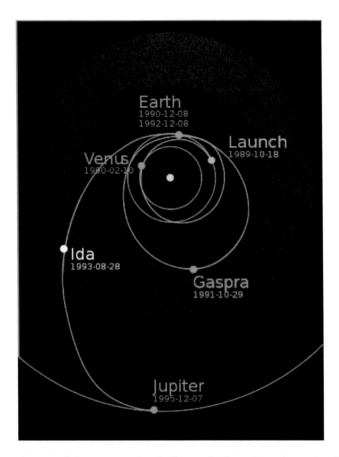

Fig. 4.5 The trajectory of the space probe Galileo to Jupiter where the probe dipped into the atmosphere. On the way to Jupiter the two minor planets Ida and Gaspra were visited. (NASA)

distance to the Moon (1,549,719 km). The asteroid was visible in the southern sky with very good visibility with binoculars. The penultimate approach (distance 7,524,773 km) took place on November 9, 2008. The last approach took place on December 12, 2012.

The asteroids, of which about 600,000 are known, are very small, only 14 of those known having diameters greater than 250 km. Although they are concentrated into belts, even in the densest part of the main belt between Mars and Jupiter, the average distance between two asteroids is still more than a million km. One does not have to fly a zigzag course through this belt with a spacecraft, as already mentioned.

With regard to the chemical composition, a distinction is made:

- C asteroids; carbon-rich objects
- S-asteroids: consist mainly of silicate compounds,
- M asteroids: consist mainly of metals.

Fig. 4.6 The asteroid Eros. (NASA)

There are studies to capture M asteroids and use them as raw material deposits. The capture of a 1-km M asteroid could cover the world's consumption of industrial metals for several decades. At present, however, such considerations are not yet feasible, also for economic reasons.

On 12 February 2001, the asteroid Eros was visited by a space probe (NEAR) (Fig. 4.6). This asteroid comes relatively close to the Earth and was previously used for distance determinations. From earth, the position of the object has been determined from two observatories as far away as possible, giving the parallax. Eros has an extension of $33 \times 13 \times 13$ km and belongs to the class of S asteroids. Its mean density is 2.4 g/cm^3. The closest point of its orbit to the Sun is 1.13 AU, the furthest point to the Sun is 1.78 AU.

In 2005, a Japanese space probe took samples from the asteroid Itokawa, and in 2010 a capsule containing these samples landed on Earth. In 2007, the Dawn spacecraft was launched. Dawn flew past Mars in 2009 (Gravity Assist), putting it on orbit to Vesta, where it arrived on July 16, 2011. During this process, the ion propulsion system was in operation for almost 70% of the entire journey time. It consumed about 250 mg of xenon per day, with a total of 450 kg of xenon in the tank. In 2015, it visited the dwarf planet Ceres (Fig. 4.7).

4.2.2 Trojans

This is a group of asteroids at Lagrange points L4 and L5 in the Sun-Jupiter system. At these points, the gravitational forces of the Sun and Jupiter cancel each other, and objects can remain stable near these positions for long periods of time. L4 and L5, respectively, form an equilateral triangle with the Sun and Jupiter. This is sketched in Fig. 4.8. The

Fig. 4.7 Image of the dwarf planet Ceres from a distance of 13,500 km. (NASA/JPL-Caltech/R. Hurt)

Fig. 4.8 The group of Trojan asteroids is located at points L4 and L5, respectively, in the Sun-Jupiter system

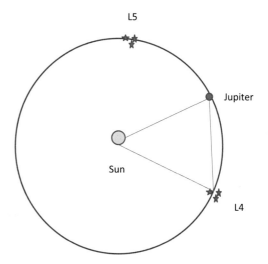

equilateral triangle is indicated only at the bottom. Since the major semi-axis of Jupiter is 5.2 AU, the lengths of the triangle are 5.2 AU each.

The Lagrange points L4 and L5 are equilibrium points which always occur when dealing with a *restricted three-body problem*. One considers two masses (in the case of the Sun and Jupiter) and investigates the motion of a third mass (in this case a Trojan asteroid). The points L4 and L5 are stable points, if one brings a not too big disturbance to a body located there, it returns to the point.

4.2.3   Earth Orbit Cruisers, Are We in Danger?

About 200 Earth-crossing asteroids are known, but there are probably 2000 larger than 1 km. Such objects can, of course, collide with terrestrial planets or be accelerated by a close encounter so that they leave the solar system. About 1/3 will collide with Earth at some point, with a major collision expected about every 100 million years. 65 million years ago, the dinosaurs became extinct when Earth collided with an asteroid only 10 km across. So an object of that size can wipe out almost all animal and plant life on Earth. During the last 500 million years, there have been at least five episodes of such *mass extinctions.* In the case of the event 65 million years ago (also known as the K-T event), the impact crater is known; it was accidentally found during oil drilling near Mexico's Yucatan Peninsula. This *Chicxulub crater* has a diameter of about 180 km. Figure 4.9 shows the location of the crater near what is now the Yucatan Peninsula. The crater has been buried by layers of sediment up to 1000 m high over the past 65 million years. The diameter of the impact object was about 10 km, which can be calculated from the mass of sedimentary layers from this time. On the right of the figure you can see an enlarged section and the gravity anomaly measurements shown clearly indicate the crater.

In Europe, the *Ries crater* is found in the Swabian Alb. Its location is shown in Fig. 4.10. The Steinheim Basin was formed by sediments ejected during the impact. It was formed 15 million years ago and its diameter is 24 km. The asteroid had a diameter of about 1.5 km. The speed when it entered the Earth's atmosphere was 20 km/s = 72,000 km/h. *Moldavites* formed from the material knocked out on impact, which was transported up to 450 km. In contrast to the K-T event (K-T stands for Cretaceous-Tertiary), where there was a mass

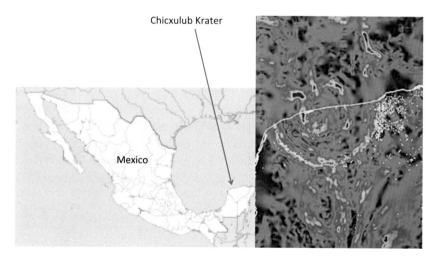

Fig. 4.9 The location of Chicxulub crater in Mexico. (Wikimedia Commons; cc-by-sa 3.0)

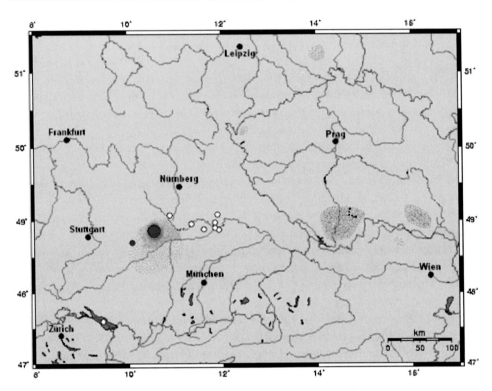

Fig. 4.10 The location of the Ries crater, the Steinheim Basin and the sites of the moldavites (green). (Creative Commons; cc-by-sa 3.0)

extinction of animal and plant species, such an event is not detectable in the formation of the Ries crater (Fig. 4.11).

The last major impact occurred in 1908 in the Siberian *Tunguska* (actually a small river). Travelers on the Trans-Siberian Railway reported suddenly seeing a bright fireball in the daytime sky. It was not until years after the event that a scientific expedition was undertaken to the affected area, and all the tree trunks were found lying in a radial direction over an area equivalent to the size of the German state of Saarland, which can only be explained by the blast wave from an explosion. The image shown in Fig. 4.12 dates from 1927, and today a small lake is found there, but hardly any remains of a crater or the impacted object. The diameter of the asteroid may have been between 30 and 80 m and it must have exploded between 5 and 14 km above the ground. Since hardly any remains were found, it must have consisted mainly of ice.

Fig. 4.11 Nördlinger Ries with Steinheimer Feld *(small circle)* on a satellite image. (NASA)

Do asteroid impacts pose a real threat to life on Earth? One thing is certain: there have been mass extinctions in the animal and plant kingdoms, and some of these events may have been caused by an asteroid impact. Earth-crossing asteroids are sought or monitored in many observing programs. There is the Turin scale for their hazardousness. The scale ranges from green to red, where a collision with Earth is then considered very likely. Still, the threat of being killed by an impact is many orders of magnitude less than being killed in a plane crash or even a car accident. Even the probability of winning the lottery is many times greater.

▸ Asteroids are objects usually well under 100 km in diameter; some cross the Earth's orbit and can also impact; the last major impact was 65 million years ago and wiped out 80% of life on Earth. In the future, near-Earth asteroids could also become interesting as suppliers of raw materials.

Fig. 4.12 Photograph of the area around Tunguska from 1927. (Wikimedia)

4.3   Comets

4.3.1   Help, the Comet Is Coming

Comets since ancient times were considered to bring bad luck. The reason for this is quite simple. They do not follow the rules of the sky, suddenly appear somewhere and disappear again. Their appearance is also unusual: they appear diffuse and usually form a spectacular long tail, always directed away from the sun. According to the laws of Kepler and Newton, it was possible to calculate the orbits of the planets exactly. In the eighteenth century, the general view was that everything was calculable, predeterminable. The cosmos resembled an exact clockwork. The comets therefore fit all the less into these ideas. They move somewhere in the sky, not along the ecliptic, or at least close to it, like the other planets and the moon.

The scene of the birth of Christ with a comet ("Star of Bethlehem") transmitted in many depictions most likely does not show a comet, but at that time it was a close encounter of the planets Jupiter and Saturn. This was known to astronomers of the time, and so the three magi from the East (the Magi, in reality they were astrologers) set out to find the baby Jesus.

With telescopes you can see about 20 comets a year, and every few years a comet becomes spectacularly visible to the naked eye.

4.3.2 Periodic Comets

E. Halley (1656–1742) found in 1705 that the comet apparitions of 1531 and 1607 (observed by Kepler) and 1682 must be the same object, and he predicted the comet's return in 1759. As Halley had already died in 1742, he did not live to see the triumph of his prediction. The comet named after him thus has an orbital period around the sun of between 75 and 77 years. Its farthest point from the Sun is more than 35 AU, or 35 times the distance Earth-Sun, and its nearest point from the Sun is 0.58 AU, or just over half the distance Earth-Sun. Many periodic comets have periods of less than 20 years. Comet Halley was last seen in 1985/1986. Already on Babylonian cuneiform tablets reports about appearances of Halley's comet are given.

The Italian painter Giotto di Bondone was deeply impressed by Halley's apparition in 1301 and painted it in his famous fresco "Adoration of the Magi" (Fig. 4.13). Since that time, the Star of Bethlehem has been depicted as a comet. The next apparition of Comet Halley is expected in 2061.

One finds many comets whose farthest point from the Sun, aphelion, is near Jupiter's orbit. This so-called *Jupiter family of* comets comes about because of Jupiter's influence on comet orbits. Jupiter's gravity deflects comets and so comets with extremely long orbital periods around the Sun become short-period comets. Orbital periods, and thus the times at which comets recur, range from five to eleven years. The other large planets also have comet families.

4.3.3 What Are Comets?

The sight of a bright comet in a telescope is disappointing for laymen, one sees only a diffusely shining bright cloud. The best way to see bright comets is with the naked eye or with binoculars. Comets consist of:

- Core: irregularly shaped, some 10 km in size. It consists of rock and ice. On approach to the Sun, when the comet is within the orbit of Mars, the volatile components evaporate. The astronomer Whipple (1906–2004) refers to it as a "dirty snowball".
- Coma: formed from the evaporating material. Can reach earth size. Consists of water and e.g. CO_2. The UV radiation of the sun splits water molecules and a huge hydrogen cloud is formed.

Fig. 4.13 The Adoration of the Magi; fresco by the Ital. Painter Giotto, which shows Halley's comet. Capella degli Scrovegni, Padua. (© picture alliance/akg images/Cameraphoto)

- Tail: Comet tails always point away from the Sun; a distinction is made between dust tails and ion tails.
- Dust tail: directed away from the sun by radiation pressure. Light consists of photons which exert a momentum (= pressure).
- Ion tail: The ion tail is called plasma tail. It appears bluish and is pushed away directly by the solar wind. The solar wind is a stream of charged particles emitted by the Sun.

The ion tail is long and narrow, the dust tail appears broad and often curved. Particles that are further away from the sun move slower around it than particles that are closer.

Fig. 4.14 Comet Neowise, which was clearly visible in summer 2020. (A. Hanslmeier, Pretal Private Observatory)

In 1997 one could observe the comet Hale-Bopp with the naked eye for several months. Its core diameter is 50 km, which is about three times larger than Halley's comet. An amateur astronomer thought he saw an unknown star near the comet on photographs. This prompted members of the *Heaven's Gate* cult to commit a collective suicide. In 2020, Comet Neowise was clearly visible in the summer (Fig. 4.14).

Comet Wild 2 was studied by the Stardust spacecraft in 2004, and samples of the comet's coma were brought to Earth (Fig. 4.15). The nucleus of the comet measures only 5 km. The average density is very low: 0.5 g/cm^3.

The Rosetta mission started in March 2005. After the flybys of the two asteroids Steins and Lutetia, the probe was put into deep sleep, all systems were switched to standby. In January 2014, the probe was awakened and in August 2014 it became an artificial moon of comet Churyumov-Gerasimenko. The highlight of this mission was the successful landing of the Philae lander on the surface of this comet on November 12, 2014. Unfortunately, the lander did not touch down horizontally as planned and therefore the power supply of the solar cells was not sufficient for prolonged radio contact. The mission was directed by the European Space Operations Centre ESOC in Darmstadt, but the runtime of the communication signals was 30 min due to the distance of the probe. Figure 4.16 shows the completely irregularly shaped comet.

Fig. 4.15 The nucleus of comet Wild 2. (NASA)

Fig. 4.16 The comet Churyumov-Gerasimenko. (NASA/ESA)

4.3.4 Comet Impacts

Comets lose mass during the outgassing processes. No one can say exactly how much, and this makes predicting comet orbits difficult. After a few dozen orbits, they may have lost most of their volatile components, and then they move around the Sun as "normal" asteroids.

Comets, when they get close to a planet, can also break up, that is, be torn apart by tidal forces. This happened with the comet Shoemaker Levy. In 1992, the comet passed Jupiter within the Roche boundary. This is the area where a body is torn apart by tidal forces. Shoemaker Levy broke into 21 fragments between 500 and 1000 m in diameter. These lined up on a chain several million km long. Between July 16 and 22, 1994, the comet fragments then collided with Jupiter. They plunged into Jupiter's atmosphere at a speed of 60 km/s, and the explosive force developed in the process was equivalent to about 50 million Hiroshima bombs (the Hiroshima atomic bomb had an explosive force of 13 kT TNT). The impacts left dark spots on Jupiter up to 12,000 km in size, which were visible for several months (Fig. 4.17).

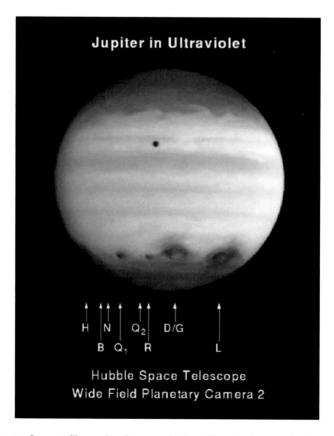

Fig. 4.17 Impact of comet Shoemaker Levy on Jupiter. The circular spot in the upper area is the shadow of one of Jupiter's moons. The image was taken in UV light. Hubble Space Telescope

4.3.5   Where Do Comets Come from?

Comets originate from the Oort cloud, which envelops the solar system. Random disturbances cause them to enter the interior of the solar system. There they can be deflected by the large planets, and so long-period comets become short-period ones. The investigation of comets is particularly interesting, because it concerns here unchanged material from the early time of the solar system. By comet impacts also a large part of the water could have come on earth. By the way, most comets have been found with the solar satellite SOHO. Comets that come very close to the sun can be observed with it and are called "sungrazers".

Comets originate from the Oort cloud and enter the inner solar system through disturbances, where their orbit is deflected by the large planets. Gases escape from the comet's nucleus, which is only a few km in size, when it approaches the Sun, producing the spectacular comet tails.

4.4   Dwarf Planets

Since 2006, this classification of solar system objects has been used for objects similar to Pluto. Their mass is not enough to be perfectly spherical and, unlike the other planets, they have not yet completely cleared their orbits.

4.4.1   Pluto

In February 1930 this object was found by Tombaugh. Pluto (Fig. 4.18) is very similar to Neptune's moon Triton. Its orbit around the Sun is strongly elliptical and therefore its solar distance can range from 4.4 to 7.3 billion km. Until February 1999 it was even inside the orbit of Neptune. Pluto needs 248 years for one orbit around the sun. For a long time its mass was not known exactly, because the disturbances Pluto exerts on Neptune are very small. Then, in 1978, Pluto's moon Charon was discovered. While Pluto itself measures a little over 2000 km in diameter, making it smaller than our moon (see Fig. 4.19), Charon is only half that size. However, this is a lot in relation to the size of Pluto and therefore it is better to speak of a double dwarf planet. Four more tiny moons were later found. Charon moves retrograde around Pluto and Pluto itself also rotates retrograde. In 2015, the New Horizons probe explored the Pluto system in more detail for the first time.

Like Neptune's moon Triton, Pluto is also a Kuiper Belt object.

Figure 4.20 shows an image of the moon Charon taken by the New Horizons space probe. Since the surface appears relatively smooth, it should also be young. A total of five moons of Pluto are known.

Fig. 4.18 The dwarf planet Pluto with its largest moon Charon. (Image: Hubble telescope)

The so-called heart of Pluto can be seen in Fig. 4.21. This formation is about 1600 km across. The image was taken in July 2015 from a distance of about 760,000 km. The few visible structures suggest geological activity.

In Fig. 4.22 one can see details on the surface of the dwarf planet Pluto. These are icebergs up to 3000 m high.

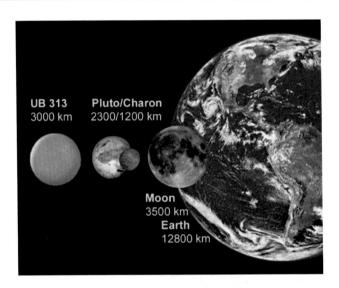

Fig. 4.19 Size comparison of two dwarf planets with the Moon and Earth. (NASA)

Fig. 4.20 The pluto moon Charon shows a relatively smooth young surface. (NASA/New Horizons)

Fig. 4.21 The heart of Pluto. (NASA/New Horizons)

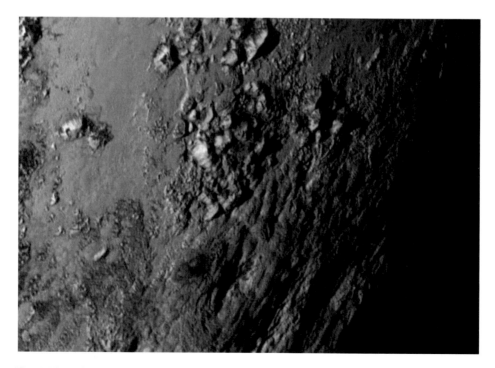

Fig. 4.22 Icebergs on Pluto's smooth surface. (NASA/New Horizons)

4.4.2   Other Dwarf Planets

Ceres (Fig. 4.7), discovered in 1801 in the asteroid belt between Mars and Jupiter, is now also listed as a dwarf planet. Other dwarf planets are the objects Sedna (~1400km, strongly eccentric orbit, perihelion distance 76 AU, apohelion distance 900 AU, orbital period around sun 10,787 years, strong reddish color), Quaoar (~1250km, major semimajor axis 43.5 AU, crystalline water ice was detected on its surface with the 8-m Subaru telescope in 2004, an indication of internal heat sources due to radioactivity), Eris (also called Xena, larger than Pluto, diameter 2400 km, perihelion distance 37.8 AU, apohelion distance 97.5 AU).

The dwarf planet Haumea (named after a Hawaiian goddess) is strongly flattened because of its rapid rotation of just under 4 h. The diameter of the pole is 1100 km: The equatorial diameter is 2300 km the pole diameter is 1100 km. A star occultation by Haumea in 2017 detected a ring about 70 km wide around it. Thus, it is the first discovered dwarf planet with a ring (Fig. 4.23). In addition, two moons are still known. The orbital period around the sun is 285 years, the perihelion distance 35 AU, the aphelion distance 51 AU. Spectroscopic measurements suggest that its surface is covered with crystalline water ice. The lifetime of crystalline water ice is only about ten million years because of cosmic rays on the unprotected surface. It is assumed that Haumea, together with its moon and ring, was formed by a collision of two dwarf planets.

▸ Pluto and other objects are counted among the dwarf planets, whose number is constantly increasing due to new observations.

Fig. 4.23 This is what the dwarf planet Haumea with its ring might look like. (R. Kayser World of Physics after J. L. Ortiz)

4.5   Meteoroids

4.5.1   Shooting Stars

Everyone has observed them: shooting stars. If you make a wish while watching them flash briefly, that wish is supposed to come true. The problem is, they only light up for a very short time.

Shooting stars are small objects of only 1–10 mm in size that penetrate the Earth's atmosphere. Scientifically, the phenomenon is called a meteor. They are burning up meteoroids; if remains are found on the surface of the earth, this is called a meteorite. Fireballs, called bolides, are about the size of tennis balls.

They are divided into the following groups:

- Fireballs: >1 cm, mass >2 g; about 1 t falls to earth per day.
- Shooting stars: 1 mm–1 cm; masses 2 mg–2 g; about 5 t fall to earth per day.
- Micrometeors: less than 1/10 mm; masses less than 0.002 g; 1000–10,000 t fall to earth daily.

The flashing occurs at an altitude of about 100 km due to air friction.

4.5.2   Meteor Streams

Meteoroids are dissolution products of comets. Whenever the earth touches the orbit of a comet during its orbit around the sun, a particularly large number of meteors are observed. All shooting stars seem to come from a certain point in the sky (radiant). This is an illusory effect. When you drive a car through heavy snowfall, you also get the impression that the snowflakes are coming from a certain spot.

Table 4.1 lists some known meteor streams. A meteor stream is named after the constellation in which the radiant is located. The best known are the *Perseids,* which

Table 4.1 Known meteor streams

Designation	Appearance	Maximum	Number per hour
Quadrantids	December 28–January 12	January 3	120
Lyrids	16 April–25 April	April 22	30
Aquarids	19 April–28 may	May 5	60
Perseids	17 July–24 august	August 12	100
Tauride	15 September–25 November	November 10	Variable
Orionids	October 2–November 7	October 21	23
Leonids	November 6–November 30	November 17	15
Geminids	December 4–December 17	December 14	120

Fig. 4.24 Meteor stream. (NASA)

originate from comet 109P/Swift Tuttle. They are also popularly known as the Tears of Laurentius. The Orionids go back to comet Halley and so do the Aquarids in May.

In Fig. 4.24 one can see a meteor stream.

The meteorite that struck near Chelyabinsk in Russia in February 2013 injured more than a thousand people (shards of glass from broken windows) from the blast wave generated by the impact. Its size was between 10 and 15 m.

▸ Shooting stars are mm to cm sized particles that light up at an altitude of about 100 km. Larger chunks fall to earth.

The Mechanics of the Sky

5

Can the motion of the earth, moon and planets be predicted with high accuracy for all times? When the law of gravity was known, it was believed that the motion of the planets, the moon and the sun in the sky could be calculated exactly, provided that all the initial data were known precisely enough. The universe was therefore deterministic, predictable. P. S. Laplace (1749–1827) said that if one knew the exact position and speed of all objects in the universe, as well as the forces acting between these bodies, then the future would be completely predictable. A being that knew these quantities was called a Laplace demon – today we would call it a supercomputer. Is the universe completely predictable, are planetary orbits stable for all time? There were even prizes offered for answering this question.

In this section you will learn

- which is a two-body problem,
- what the Lagrange points mean,
- how solar and lunar eclipses can be predicted,
- how stable the solar system really is.

5.1 The Lunar Orbit

As already indicated, the moon moves on an elliptical orbit around the earth. It takes 27.33 days to orbit the earth. After one sidereal month the moon can be found again at the same place in the starry sky. However, it does not have the same phase, because the earth has moved a bit along its orbit around the sun during this time. The synodic month

© The Author(s), under exclusive license to Springer-Verlag GmbH, DE, part of
Springer Nature 2023
A. Hanslmeier, *Fascination Astronomy*,
https://doi.org/10.1007/978-3-662-66020-1_5

Fig. 5.1 The moon's orbit inclined to the plane of the earth's orbit (ecliptic)

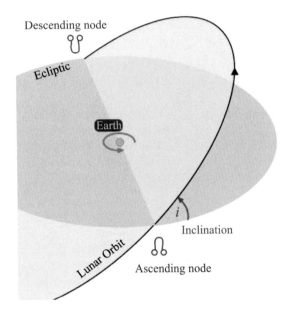

describes the return to the same phase of the moon and the length is 29.53 days. If the moon is full today at 12 o'clock, then the moon will be full again in 29.53 days.

The Course of the Moon The Moon moves in an orbit around the Earth that is inclined at about 5.2° to the Earth's orbital plane (ecliptic). The lunar orbital plane intersects the ecliptic plane at two points called nodal points (Fig. 5.1). Since the lunar orbital plane is inclined, a torque is applied to it by the Sun and the planets. Let us imagine a spinning top that we set in rotation. If we push this gyroscope, then it reacts, as shown in Fig. 5.2, with a wobbling motion, which is called precession.

The same effect can be observed at the moon's orbit, which can be thought of as a kind of gyroscope, the mass of the moon can be thought of as distributed along the moon's orbit. That is why the length of the draconic month, i.e. the period of time that passes until the moon returns from a position to a certain node, is only 27.212 days. So if the Moon is at one of the nodes today, 27.212 days will pass before it is there again.

The formation of the phases of the moon is illustrated in Fig. 5.3.

- New Moon: Moon is between Sun and Earth,
- Full Moon: Moon is opposite the Sun in the sky (opposition), it rises in the East when the Sun sets in the West or the Moon sets in the West when the Sun rises in the East.
- First or last quarter: Seen from the earth, the angle between the sun and the moon is 90°.

Since the Moon's orbit around the Earth is elliptical, its distance from the Earth changes:

Fig. 5.2 Imagine a rotating
gyroscope to which a force is
applied, then observe the
precession

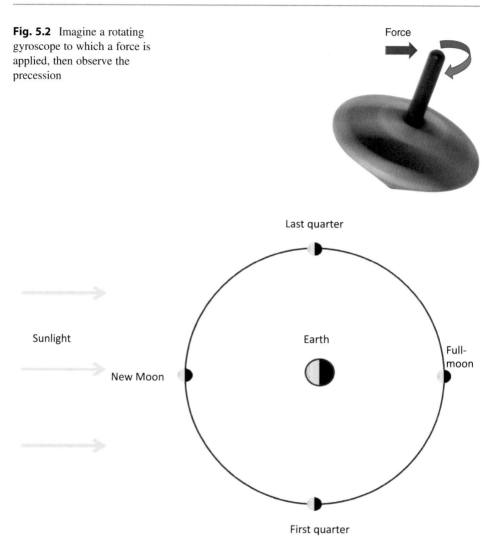

Fig. 5.3 On the origin of the moon phases

- The closest point to the earth is called perigee (the calendar usually says: "moon near earth"), distance to earth 362,000 km,
- and the farthest point from earth as apogee (the calendar says "moon in earth distance"), distance to earth 405,000 km.

When the moon is near the earth at the same time during the full moon phase, it shines a little brighter, which is called a supermoon. In this case, the moon appears about 14% larger and 30% brighter. For science this has no meaning, as well as Blue Moon. One speaks of a Blue Moon when there are two full moons within a calendar month (the second full moon is then the Blue Moon). This happens every 2.5 years.

Fig. 5.4 Solar eclipses occur
when the arth is in the shadow of
the moon

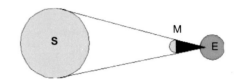

As we know from physics, two bodies move around their common centre of gravity. Exactly the same with the system earth-moon. The moon does not move around the center of the earth, but around the center of gravity (barycenter) of the system earth-moon. The mass of the moon is 1/81 of the mass of the earth. The average distance earth-moon is 384,400 km, so the center of gravity is located 384.400 km/81~4670 km from the earth center, i.e. about 1700 km deep in the earth mantle.

Influences of the moon on the earth are, apart from the origin of the tides, not provable.

5.1.1   Solar and Lunar Eclipses

In the sky, the moon appears to us about 0.5° in diameter, which is about the same as the diameter of the sun. Solar eclipses occur when the moon moves between the earth and the sun, which is only possible during the new moon phase. For the occurrence of a solar eclipse (Fig. 5.4) the following points must be fulfilled:

- New Moon,
- Moon is on its orbit in the ecliptic plane, i.e. near one of the two intersections of the Moon's orbit with the ecliptic.
- When the moon is close to the earth during a solar eclipse, it appears larger and the eclipse lasts a little longer.
- When the Earth is near the farthest point of its orbit from the Sun (early July), the eclipse also lasts longer because the Sun then appears slightly smaller.
- ▸ Eclipses only occur when the moon is close to the ecliptic, at full moon (lunar eclipse) or new moon (solar eclipse).

The points of intersection of the moon's orbit with the ecliptic are called *nodal points*. *They* move around the ecliptic with a period of about 18 years, so solar eclipses repeat every 18 years. Since the shadow of the moon on the earth is small, the totality zone of an eclipse on the earth is only about 200 km, and if the moon is at far a distance from the earth during the eclipse, it no longer completely covers the sun, and an annular eclipse is observed. Outside the totality zone one sees a partial solar eclipse. Although the eclipses repeat with a period of about 18 years, the places on earth where you can observe the eclipse shift westwards. This was already known to the Babylonians *(Saros cycle)*. The 18-year period is related to the movement of the lunar nodes (intersections of the lunar orbit and ecliptic) (Fig. 5.5). The Earth's shadow is much larger than the Moon's shadow, so we see lunar eclipses from all locations on Earth where the Moon is above the horizon during

Fig. 5.5 Lunar eclipses occur
when the moon is in the shadow
of the earth

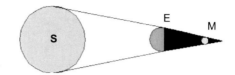

the eclipse. So for a given location on Earth, there are many more lunar eclipses than solar
eclipses.

There are also partial lunar eclipses, where the moon does not enter the earth's umbra,
but only its penumbra. But these penumbral eclipses are only visible to the experienced
observer. During a total lunar eclipse the moon does not disappear completely, but usually
appears dark red faintly shining in the sky (also called blood moon). This illumination is
caused by sunlight scattered in the Earth's higher atmosphere. In the past, the darkness of
the moon during a total lunar eclipse was used to infer the state of the Earth's higher
atmosphere.

5.2   Earth Axis and Gyroscope

5.2.1   Precession

We have already pointed out the analogy to a pushed gyroscope, which reacts with a
looping movement, when discussing the moon's orbit inclined to the earth's orbit plane (=
ecliptic). A similar movement is also shown by the earth axis, more exactly its direction.
The Earth's axis is inclined to the Earth's orbit by about 23.5° from normal (see also
Fig. 5.6, angle ε). The Sun, Moon, and planets try to make the Earth's axis equal to the
normal to the Earth's orbital plane, the Earth reacts like a gyroscope, and the Earth's axis
makes a wobbling motion called precession (Fig. 5.7). The period of this motion is about

Fig. 5.6 Celestial equator and
vernal equinox. The angle ε is
currently about 23.5°.
(Wikimedia Commons; S.fonsi;
cc-by-sa 3.0)

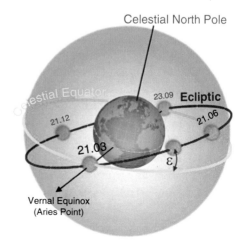

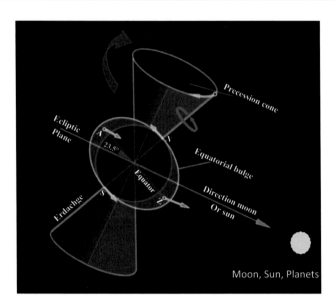

Fig. 5.7 The Sun and Moon exert a torque on the Earth's equatorial bulge, and the Earth responds by moving in precession. (H. C. Geier)

26,000 years, which is sometimes called the Platonic year. Does this have any practical implications? The answer is yes and no. At present, in the northern hemisphere of the Earth, we live in a fortunate situation: the Earth's axis points pretty much in the direction of a second magnitude star, which is not one of the brightest stars in the sky, but there are no brighter stars in its vicinity: the North Star (Polaris). However, as a result of the precessional motion of the Earth's axis, this is changing over time. In about 12,000 years, the brightest star in the northern night sky, Vega, will be Polaris.

The position of the vernal equinox also shifts as a result of precession. The vernal equinox is the place where the sun is located at the beginning of spring. Then the sun is exactly at the celestial equator. Day and night are of equal length, both halves of the earth, the northern and the southern, are illuminated by the sun for the same length of time, these two points are called equinoxes. However, the location of the vernal equinox shifts as the direction of the Earth's axis changes. More than 2000 years ago, the vernal equinox was located in the constellation of Aries and then moved through the constellation of Pisces and will continue to move into the constellation of Aquarius. This is celebrated in esoteric circles as the beginning of a new age. But as I said, nothing exciting is happening and in 2000 years the vernal equinox will be in the constellation of Capricorn. Astrology still calculates as if there is no precession, so the signs of the zodiac in astrology no longer have anything to do with the actual zodiac constellations in the sky.

▶ Precession changes the position of the vernal equinox in the sky.

5.2.2   Nutation

Since the moon's orbit is inclined by 5° and the position of the moon's nodal points shifts with a period of about 18 years, this also exerts a force on the earth's axis. The earth reacts again as a gyroscope and this results in the nutation movement. The amplitude of this movement is much smaller and the period much shorter.

By the way, astronomy uses a coordinate system with the vernal equinox as the origin. Since its position shifts, one must always indicate the so-called *equinox* when specifying the coordinates, i.e. the date to which the coordinates refer. Today one uses mostly equinox 2000.

5.3   The Annual Cycle of the Sun

As seen from Earth, the Sun appears to pass through all the constellations of the ecliptic during the course of a year. Since most of these constellations are animals, they are called the *zodiac*. The constellations of the zodiac are: Aries, Taurus, Gemini, Cancer, Leo, Virgo, Libra, Scorpio, Sagittarius, Capricorn, Aquarius, and Pisces. The position of the vernal equinox shifts due to the precessional motion of the Earth's axis and is currently located in the border area of the constellations Pisces-Aquarius.

- In summer, the sun is high in the sky in the northern hemisphere in the constellations of Taurus, Gemini, Cancer,
- in winter it stands in the northern hemisphere low in the sky in the constellations Scorpio and Sagittarius.

The apparent path of the sun is called the ecliptic, the word means eclipse line. It was recognized that eclipses always occur when the moon is close to the ecliptic at full moon or new moon (nodal point).

The ecliptic is inclined to the equatorial plane of the earth by 23.5°. This means that at the beginning of summer the sun is 23.5° north of the celestial equator (this is the projection of the earth's equator on the celestial sphere), but at the beginning of winter it is 23.5° south of it.

The height of the *celestial* equator above the observer's horizon depends on where this observer is located on Earth, i.e. on his latitude. At the equator, the celestial equator passes through the zenith, i.e. directly vertically above the observer; at the Earth's pole, the celestial equator is exactly at the horizon. Therefore, the altitude of the celestial equator:

$$h = 90 - \phi. \tag{5.1}$$

ϕ is the latitude. This also indicates the altitude of the celestial pole. Let's say we are at the North Pole of the Earth, then the Pole Star is exactly perpendicular above us and the latitude is of course $\phi = 90°$. At the Earth's equator, the altitude of Polaris is zero and $\phi = 0°$.

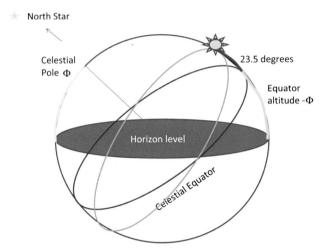

Fig. 5.8 Sun at noon at the beginning of summer; celestial equator *(red)*, solar orbit (ecliptic, *green*) are inclined to each other by 23.5°

▸ Thus, the altitude of Polaris above the horizon is equal to the latitude ϕ.

With this we can easily calculate how high the sun is at noon at the beginning of summer at a location of the geographical latitude $\phi = 50°$: The sun is 23.5° north of the celestial equator on this date. The altitude of the celestial equator is $90 - 50 = 40°$. Thus, the altitude of the sun above the horizon is $40 + 23.5 = 63.5°$.

Figure 5.8 shows the position of the sun at noon at the beginning of summer. Red means celestial equator, green the ecliptic.

Also the maximum height of the moon can be calculated easily, its orbit is inclined about 5° against the ecliptic. So the maximum possible height of the moon is $90 - \phi + 23.5 + 5°$. The full moon is high in the sky in winter and very low in summer. However, it is opposite the sun, i.e. in opposition, i.e. at full moon it rises when the sun sets and sets when the sun rises.

The seasons are caused by the difference in solar radiation due to the 23.5° tilt of the Earth's axis. For the northern hemisphere, the days from the beginning of spring to the beginning of autumn are longer than twelve hours and the sun's rays fall more steeply, resulting in greater warming. For the southern hemisphere, the situation is just the opposite. So seasons have nothing to do with the fact that the Earth's orbit is a slight ellipse. Climate changes in the past (ice ages) are related to the fact that the following variables are variable due to perturbances from the other planets:

• Inclination of the Earth's axis: is currently 23.5°, can vary by a few degrees around this value; the greater the inclination, the more pronounced the seasons. Planets with an axis inclination close to zero show no seasonal effects (e.g. Jupiter).

- Eccentricity of the Earth's orbit: The differences between the vicinity of the Sun (perihelion) and the distance from the Sun (aphelion) can increase, so the Earth's orbit becomes more "elliptical".

These two disturbances have periods of several 10,000 years, but if the periods overlap accordingly, this can trigger *ice* ages (Milankovic's theory).

Our Earth's axis is stabilized by the Moon, this is not the case with Mars. The inclination of the Martian rotation axis can vary irregularly between zero and 60°. This explains the strong climate changes suspected there.

5.4 Planetary Orbits

5.4.1 The Kepler Laws

Kepler already described the planetary orbits in his three famous laws: The first law states that planets move on an ellipse, in one focal point of which is the Sun. He came to this law through his exact calculation of the orbit of Mars, which has a certain eccentricity (e.g. the orbit of Jupiter is almost circular). Kepler was given the task of accurately calculating the orbit of Mars by his superior, Tycho Brahe, who also made very accurate observations. At the time, it was postulated that planetary orbits had to be circular, since only the circle represented some sort of perfect motion. Therefore, the description of planetary orbits as ellipses was truly revolutionary and met with opposition.

From Fig. 5.9 one can see:

- Planet's distance from the Sun at perihelion: $r_P = a - ae = a(1 - e)$.
- Planet's distance from the Sun at aphelion: $r_A = a + ae = a(1 + e)$.

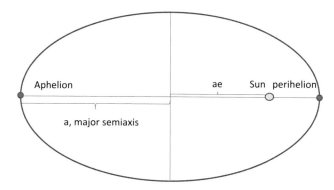

Fig. 5.9 Planetary orbit as an ellipse. At perihelion the planet is closest to the Sun, at aphelion it is farthest away

Fig. 5.10 The epicycle theory
to explain the planetary loops in
the sky

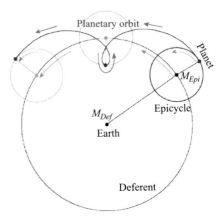

Planetary orbits used to be described by the complicated epicycle system (Fig. 5.10), which can already be found in Ptolemy's work (c. 100 to c. 160 AD). Thus, the loops of the planets in the sky could be explained, albeit in a complicated way. According to the *epicycle theory,* the planets move on epicycles, i.e. circles that move on a great circle that moves around the sun. In principle, this allows us to describe the orbits of the planets very accurately, but many epicycles are needed to describe them accurately. So the description becomes very complex. One of the basic rules of physics is: the simpler, the more correct. Nature is simple, all theories that are very complicated were wrong, just like the epicycle theory. If you put the sun in the center and all planets move around the sun, then the looping motions can be explained quite simply. Outer planets always loop in the sky when the Earth overtakes the outer planet due to its greater orbital velocity (see Fig. 5.11).

▶ The looping motions of the planets can be easily explained by the heliocentric model.

Kepler's Second Law states that a planet moves faster around the sun when it is close to it than when it is far from it. We have explained this before. Near the Sun, a planet experiences a stronger attraction from the Sun, so it must move faster. The practical consequence for us: In the northern hemisphere of the Earth, the summer half-year lasts somewhat longer, since at present the point of the Earth's orbit furthest from the Sun is reached at the beginning of July.

Kepler's Third Law is very commonly used and has great significance. The exact form is:

$$\frac{a^3}{T^2} = \frac{G}{4\pi^2}(M_1 + M_2). \tag{5.2}$$

Here we consider the motion of two masses M_1, M_2, where M_1 is at the focal point of the orbital ellipse of M_2, a is the major semi-axis of the orbital ellipse of M_2, T is the orbital period of M_2. Using this law, we can determine the mass of the Earth from the motion of a satellite around the Earth, for example, or the mass of the Sun from the motion of the Earth

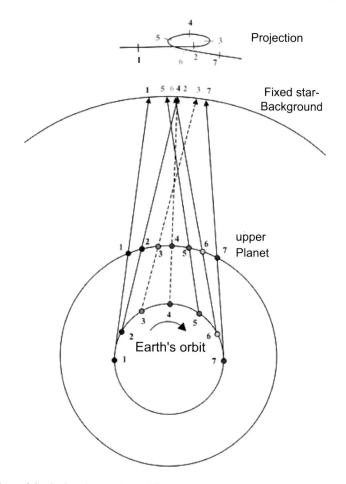

Fig. 5.11 To explain the looping motion of the outer planets. (CC-Lic.)

around the Sun, or estimate the mass of the Milky Way from the motion of the Sun around the center of the Milky Way. Let's estimate the mass of the Earth: Let a satellite move 200 km above the Earth's surface, so its orbital ellipse is equal to the Earth's radius + 200 km. The orbital period of the satellite around the earth is 90 min. The mass of the satellite, M_2, can be set to zero, since it is negligible compared to the mass of the earth. By substituting the following for the Earth's mass: $M = 6 \times 10^{24}$ kg.

We will return to Kepler's Third Law several times.

Kepler's laws are not really new laws of nature but can be derived from the known conservation laws of physics (conservation of energy, conservation of angular momentum,...).[1]

▸ The Third Kepler Law allows the calculation of masses of bodies.

---

[1] See: Introduction to Astronomy and Astrophysics, Hanslmeier, Spektrum Verlag.

5.4.2   Stability of the Planetary Orbits

Various studies show: The solar system is about 4.6 billion years old. How do the planetary orbits behave over time? Is our solar system stable? One can describe the motion of the planets and other celestial bodies by Newton's law of gravity. However, an exact solution is only possible for the motion of two bodies. For example, we can solve the "two-body problem" Earth-Sun exactly, but as soon as the effect of the Moon is added, there are only approximate solutions, which are very exact due to modern computer methods, but not exact. The so-called *N-body problem* cannot be solved analytically as soon as more than two bodies are involved.

Nevertheless, so-called long-term integrations have been carried out, i.e. solutions for the motion of the planets in the solar system over long periods of time have been sought. This results in errors:

- Computers calculate with finite accuracy.
- The N-body problem can only be solved approximately.

These errors add up. Computers always give results, but you have to check very carefully whether these results can be correct.

Long-term calculations have shown: The planetary system quite stable. The orbits of the major planets change the least. Mars, Earth and Venus also remain on stable orbits. Mercury, however, will leave us within the next two billion years. It is subject to relatively strong influences from the Sun and Venus. With respect to small bodies (asteroids, comets), our solar system is anything but stable. Again and again such objects get close to one of the large planets and their orbits are deflected or the object collides with a planet.

Many stars are binary stars, and there too it is possible to investigate whether there are stable planetary orbits. A planet in a binary system can be on a stable orbit in three areas:

- near star 1,
- close to star 2,
- in a wide orbit around the two stars.

These considerations also lead to the Lagrange points. Let us consider two masses, e.g. M_2 be the mass of the sun, M_1 the mass of the earth. Then there are five stable points where a third body can reside, but its mass should be so small that it itself does not affect the motions of the other two bodies. This is also called the restricted three-body problem. The points L_4, L_5 describe an equilateral triangle with the Earth and the Sun and are located "in front" and "behind" the Earth, respectively. In Fig. 5.12 the five points are drawn. L_1 is located between the two masses, in our case about 1.5 million km from the Earth between the Earth and the Sun, L_2 is located behind the Earth, L_3 behind the Sun. Only the points L_4, L_5 are stable.

The Lagrange points in the Earth-Sun system are very interesting for space travel. If one brings a satellite to the point L_1, then one can observe the sun there undisturbed. Figure 5.13

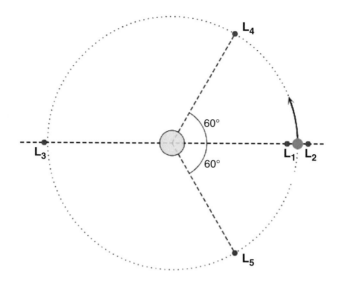

Fig. 5.12 The Lagrange points in the Earth-Sun system

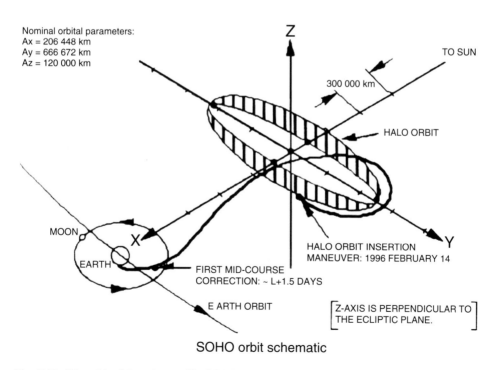

Fig. 5.13 The orbit of the solar satellite SOHO around the Lagrange point L_1. (SOHO/ESA/NASA)

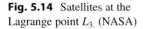

Fig. 5.14 Satellites at the
Lagrange point L_3 (NASA)

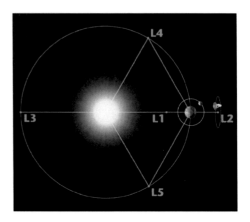

shows the orbit of the solar satellite SOHO (Solar and Heliospheric Observatory). The satellite moves around the Lagrange point L_1, which is located at a distance of about 1.5 million km from the Earth. The satellite was launched on December 2, 1995 and is an ESA/NASA mission.

   If one wants to observe the night sky, the Lagrange point L_2 is particularly suitable. Several satellites are located there: WMAP and PLANCK we have already discussed in the first chapter. The successor to the Hubble Space Telescope, the James Webb Telescope, is also positioned there. A sketch is given in Fig. 5.14. This telescope was launched at the end of 2021 and has a mirror diameter of 6.5 m (Hubble telescope only 2.4 m). The mirror consists of 18 individual segments.

   By the way, in science fiction novels a planet X appears again and again, which is supposed to be located at the Lagrange point L_3, which is invisible to us. However, we know an answer to this: an object at this point will not stay there for all time; all planets interfere with each other and therefore it would require an ongoing position correction to keep a stable planet at L_3.

   We have already discussed the asteroids at points L_4, L_5 in the Sun-Jupiter system (Trojans).

   ▶ Exactly, you can only solve the two-body problem. The solar system is stable over long periods of time. Lagrange points are special solutions of the restricted three-body problem, only L_4 and L_5 are stable.

The Sun: The Star from Which We Live

6

Without the sun, no light, no heat, no life on Earth. The earth, if it existed at all, would be a dead dark planet covered with ice. The importance of the sun for life on earth was recognized by all peoples, and in many religions the sun was worshipped as a deity. In modern astrophysics, there are two particular reasons to look more closely at the study of our nearest star. First, the Sun is the only star where we can observe a wide variety of surface details such as spots, prominences, flares, eruptions, and so on. Even with the largest telescope in the world, stars can only be seen in as pointlike spots. So the nearest star is the Sun 150,000,000 km away, Proxima Centauri is more than 300,000 times that distance. Second, we know that the Sun has a major impact on Earth and near-Earth space. During strong solar storms, satellites can become unstable, radio communications are disrupted, etc. To minimize the damage, we want to predict the *space weather, which* is mainly influenced by the sun.

In this chapter you will learn about

- the sun as the next star,
- why the sun shines,
- what sunspots are,
- whether the sun is shining constantly,
- how solar activity threatens our high-tech society.

A. Hanslmeier, *Fascination Astronomy*,
https://doi.org/10.1007/978-3-662-66020-1_6

6.1   The Sun: Basic Data

6.1.1   The Sun from the Earth

The sun is about 150,000,000 km away from the earth, this distance is used as a unit of measurement especially in the description of planetary orbits and is called "astronomical unit". On earth we receive a radiation power of about 1.36 kW per square meter. However, this only applies when the sun's rays are perpendicular and there is no absorption from the atmosphere. Therefore, plan your thermal solar power station or your photovoltaic system accordingly more generously (e.g. one takes approx. 1/5 of the value).

Seen from Earth, the diameter of the Sun's disk in the sky is about 0.5°. Coincidentally, our moon appears about the same size so that it can completely cover the sun during a total solar eclipse. However, our Moon moves away from the Earth by about 1 cm per year as a result of tidal friction. The rotation of the earth slows down, the angular momentum of the whole system must be conserved, therefore the distance of the moon increases, and in some 100,000 years there will be no more total solar eclipses, because the moon appears too small in the sky due to its greater distance.

When does the sun actually rise? This is not so easy to calculate. The light of the sun and of course of all stars is refracted by the earth's atmosphere. It comes from a vacuum (space) into a denser medium (earth's atmosphere). We see a similar effect when we look at a stick reaching halfway into water at an angle, it appears refracted. When the sun is right on the horizon, it is the image of the sun lifted by the earth's atmosphere. This effect is called *refraction* and it is 0.5° near the horizon, the closer to the zenith, the less it becomes. Figure 6.1 explains this effect. The magnitude of the refraction is r.

The magnitude of refraction depends on the state of the atmosphere, temperature, pressure, density, movement of air layers.

6.1.2   Mass and Size of the Sun

Figure 6.2 shows how to determine the ratio of the distance Earth-Sun to Earth-Moon (method of Aristarch, ca. 310 to 230 BC). If the moon is in the first or last quarter, then its distance from the sun measured in the sky is $\alpha < 90°$, seen from the earth the angle between sun and moon is exactly 90°.

Aristarchus found that the sun must be farther away than the moon, therefore larger than the moon, the correct value is 400 for the ratio, the sun is 400 times farther away than the moon, but appears the same size in the sky as the moon, consequently it must also be 400 times larger. We get the radius of the sun immediately if the distance is known (Fig. 6.3).

Fig. 6.1 The image of a star is raised by refraction in the Earth's atmosphere. The true zenith distance of the star z' is reduced by the amount of refraction r and the star appears below the zenith distance z. (Wikimedia Cosmos)

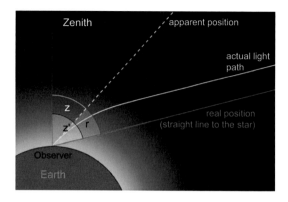

Fig. 6.2 Aristarchus' method for determining the ratio of the distance Earth-Moon to Earth-Sun

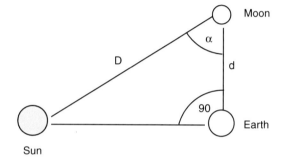

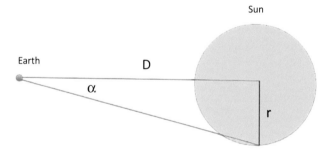

Fig. 6.3 To determine the solar radius; if the distance D is known and the apparent diameter of the sun 2α, then the true radius (diameter) follows from it

Digression From the illustration you can see:

$$\tan \alpha = \frac{r}{D} \qquad r = D \tan \alpha \tag{6.1}$$

and inserting $D = 150.000.000$ km and $\alpha = 0.5/2 = 0.25°$ gives the solar diameter of $r = R_\odot = 6.959 \times 10^8$ m ~ 696.000 km. This makes the Sun 109 times the size of the Earth and about 10 times the size of Jupiter. In 1969, humans landed for the first time on the

Moon, which is about 384,400 km away. The distance earth-moon is only 1/4 the diameter of the sun.

The Sun is 109 times larger in diameter than the Earth.

The mass of the Sun follows from Kepler's Third Law and its determination has already been discussed. In numbers: $M_\odot = 1.98 \times 10^{30}$ kg. This makes the Sun the dominant body in the solar system, all other planetary masses make up only about 0.2% of the Sun's mass. In astrophysics, the radius and mass of the Sun are used as units of measurement.

The mass of the sun is 333,000 Earth masses.

6.1.3   How Hot Is It on the Sun?

Determining the temperature of the sun requires knowledge of simple laws of physics. First, let's do a thought experiment. Let's turn on a hot plate. First, the plate becomes warm. Heat can be felt through the skin, then the plate will begin to glow faintly dark red, then bright red, and so on. Heat is physically infrared radiation, radiation with a wavelength longer than red light. The experiment shows us that the hotter the plate, the more the radiation moves to shorter wavelengths.

> Hotplate experiment: the hotter the plate, the shorter the wavelength of the emitted radiation. Heat rays are infrared rays and have a longer wavelength than red light.

The individual colours of light differ only in their wavelength. Blue light has a wavelength of about 450 nm (1 nm = nanometer $= 10^{-9}$ m), red light is over 600 nm. We see the range from about 400 to 700 nm with our eyes. At longer wavelengths, we enter the infrared range, microwave range, and finally the radio range. At wavelengths shorter than blue, we enter the ultraviolet (UV) range, X-ray range and gamma ray range.

The colors of light differ only by wavelength. Red light has a longer wavelength than blue light.

X-ray light differs from green light only by the wavelength! The entire wavelength range from X-rays to UV rays, visible light, infrared (IR), etc. is called the *electromagnetic spectrum* (see Fig. 6.4). Our eyes therefore only see a small section of the electromagnetic spectrum.

The fact that the hotter a body, the more the maximum of its radiation moves to shorter wavelengths is called Wien's law. Figure 6.5 shows the radiation curves for three different temperatures.

Digression The formula for Wien's law is:

$$T\lambda_{max} = 2.897 \times 10^{-3}\,\text{m}. \tag{6.2}$$

Fig. 6.4 The electromagnetic spectrum. The visible range with adjacent UV or IR is zoomed out. (Horst Frank/Phrood/Anony GFDL)

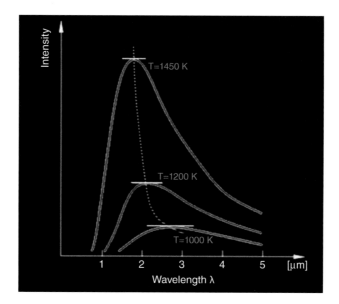

Fig. 6.5 Wien's law. The hotter the body, the more the maximum of its radiation moves to shorter wavelengths

Consider a glowing body at a temperature of 1000 K. The maximum of its radiation is then at

$$\lambda_{max} = 2.89 \times 10^{-3}/10^3 \mathrm{m} = 2.89 \times 10^{-6}\,\mathrm{m} = 2897\,\mathrm{nm}. \tag{6.3}$$

The maximum of the radiation lies far in the infrared.

Our sun has a temperature of 6000 K. From the above calculation follows:

$$\lambda_{max} = 2.89 \times 10^{-3}/6 \times 10^3 = 5 \times 10^{-7}\,m = 500\,nm. \tag{6.4}$$

The maximum of solar radiation is therefore in the green. This is also the reason why we find the colour green pleasant and why the leaves are green.

The color of a hot object (star) depends on its temperature. Wien's law says: the hotter, the more the maximum of the radiation moves to smaller wavelengths.

The temperature of the sun is therefore simply determined by measuring a radiation distribution curve, i.e. the intensity of the light at different wavelengths and then determining the wavelength at which the radiation has a maximum, i.e. at 500 nm.

The temperature of the sun is about 5800 K.

In the next chapter we will see that there are stars much hotter than our sun.

All the previously derived quantities of the Sun, mass, radius and temperature, can be determined from Earth-based observations. This is the great art in astrophysics, because the only information we get from the stars is their radiation.

6.1.4 Observations of the Sun

Already in ancient times, the sun was observed when it was low on the horizon or when its light was strongly dimmed by clouds or fog. You can observe the sun safely with a refracting telescope by *projection. A* screen is attached behind the eyepiece of the telescope onto which the image of the sun is projected. It is important never to look at the sun through a telescope, the retina would be immediately irreparably burned and the eye blinded!

Modern solar telescopes consist of an evacuated telescope tube. Vacuum does not conduct heat, the telescope does not heat up by the incident solar radiation, and one gets a very good image. Such telescopes are usually arranged in a tower shape, a mirror system directs the sunlight into the fixed telescope, and the evacuated telescope tube contains the optics (Fig. 6.6). An important quantity in solar physics is the angle one arcsecond, observed from Earth, on the Sun. At mean solar distance, the angle of one arcsecond corresponds to about 725 km. As a rough rule one can remember that a telescope in the visible range has a *resolving power* of one arc second if the telescope diameter is about 10 cm. So, under very good observing conditions, you can see details up to about 725 km in size on the Sun (in reality, the optics are usually not perfect, and the Earth's atmosphere affects the observing conditions, which can easily be seen in the trembling of the stars; this is called seeing).

The trembling of the solar image *(seeing)* is caused by turbulence in the Earth's atmosphere. Adaptive optics can significantly improve the image quality. Thin mirrors are "bent" in such a way that the image blurring is compensated for; this happens up to 100 times per second and therefore requires complex computers. Image stabilizers in cameras also work on the same principle. The world's largest solar telescope is the DKIST on Mauna Kea in Hawaii (Fig. 6.7); the telescope has a mirror diameter of four

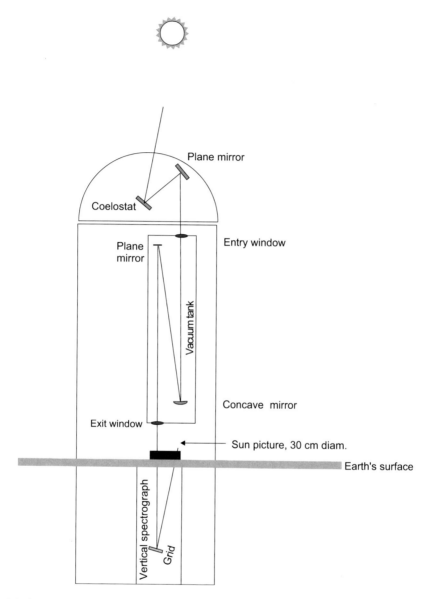

Fig. 6.6 Sketch of the vacuum tower telescope on Tenerife, which is operated by German institutes. (Leibniz Institute for Solar Physics, KIS)

meters. It is named after Daniel K. Inouye, who was US Senator for the State of Hawaii from 1963 until his death. In Europe, a telescope with a diameter of 4 m is planned (EST, European Solar Telescope).

With a mirror diameter of 1.5 mn, the GREGOR telescope on Tenerife is currently the largest solar telescope in Europe (Fig. 6.8).

Fig. 6.7 The Daniel Inoue solar telescope, DKIST in Hawaii, on the summit of Haleakala volcano

Fig. 6.8 The Gregor telescope (*left,* free-standing, dome totally down) and the vacuum-tower telescope *(right).* (KIS photo)

Fig. 6.9 The Parker Solar Probe took a picture of Venus during a swingby maneuver and discovered a bright ring, producing a night glow, in Venus' atmosphere. The streaks are produced by particles of cosmic radiation and solar wind

Observing the Sun from space has advantages and disadvantages over Earth-based telescopes:

- In space, there is no interference from the Earth's atmosphere.
- In space, you can also observe in the UV or far infrared.
- Space missions are increasingly expensive compared to Earth-based observatories.
- Space telescopes are very difficult and expensive to maintain, if at all (Fig. 6.9).

Solar Orbiter was launched on 10.2.2020 and will approach the Sun up to 60 solar radii (Fig. 6.10). Therefore, the construction of a heat shield was necessary to shield temperatures of up to 500 K. The probe will approach the Sun by means of several Venus swingbys and eventually fly over the poles of the Sun (planned around 2030).

The Parker Solar Probe approaches the Sun several times, the closest distance is said to be six million km, so it flies through the outermost layers of the solar atmosphere. During one of the Venus swingby manoeuvres, the probe discovered a night glow in the Venusian atmosphere. This occurs when oxygen atoms recombine to form oxygen molecules on the night side of Venus (Figs. 6.9 and 6.10). It was launched in 2018, and the heat shield must withstand temperatures of up to 1400 °C.

With the STEREO mission it was possible to obtain stereo images of the Sun and its outbursts by two identical probes (Fig. 6.11). This mission was launched in 2006 and in 2016 one of the two probes failed.

Fig. 6.10 Solar Orbiter will approach the Sun to within 42 million km. (ESA/NASA)

Fig. 6.11 The STEREO
mission allowed to obtain three-
dimensional images of the Sun
and its outbursts. (NASA)

6.2   The Sun: The Interior

6.2.1   Where Does the Sun Get Its Energy?

Our sun radiates enormous amounts of energy. The energy emitted by the sun in one second could meet the world's energy needs for about 300,000 years. How is this energy generated? All the energy generation processes known at the beginning of the twentieth century seemed to be far from adequate for this purpose. Let us imagine the sun as a huge pile of coal. The chemical energy released by burning it could cover the sun's current energy output for just under 10,000 years. Even then, it was clear that the solar system must be much older. Another possibility is that the Sun is shrinking, releasing gravitational energy. The process is similar to when a free-falling rock releases gravitational potential energy. But even this source of energy would dry up after a few 100,000 years. Also, the diameter of the sun would have to be slowly decreasing and this decrease would have to be measurable. So around 1900 it was completely unclear how the sun could generate these huge amounts of energy.

With the development of our ideas about the structure of atoms, the discovery of radioactivity, and quantum physics, it soon became clear that nuclear fusion was the only way to produce this energy. The Sun is made up of about 75% hydrogen, about 25% helium, and less than 1% elements heavier than helium (in astrophysics, all elements heavier than helium are called metals). Here's what happens during hydrogen fusion inside the Sun: Four hydrogen nuclei combine to form one helium nucleus (in reality, these are three different reactions). Hydrogen is the lightest and simplest element (Fig. 6.12, left). It consists of a proton in the nucleus and an electron in the shell. Inside the Sun, however, temperatures are very high. For this reason the hydrogen atom is ionized, i.e. the electron is separated and you have a mixture of protons and electrons, we say the hydrogen is ionized. Helium consists of two protons and two neutrons in the nucleus (Fig. 6.12 right) and electrons in the shell. The protons are positively charged, the neutrons are neutral, and the electrons are negatively charged. In order for two protons to fuse, the repulsive forces

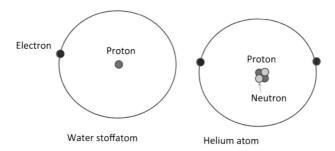

Fig. 6.12 Simplified structure of the hydrogen atom *(left)* and the helium atom *(right)*. Electrons *(red)*, protons *(blue)* and neutrons *(yellow)*

acting between two charges of the same positive charge must first be overcome (see the first chapter). Therefore, fusion only works at very high temperatures.

Physical Concept: Uncertainty Principle Even the approximately 12–15 million K inside the sun are not sufficient to overcome the repulsive forces between two protons. Only quantum physics can help here: there is the tunnelling effect. According to this effect, the protons can overcome the repulsive force with a certain probability, although, classically speaking, their energies would not be sufficient due to the temperature. The uncertainty principle states that you can never measure exactly two quantities such as location and momentum or energy and time. For example, if the location of a quantum mechanical particle is precisely determined, one knows nothing about its momentum (velocity), etc.

Digression For the uncertainty of location, momentum, energy, and time, write Δx, Δp, ΔE, and Δt, respectively, and Heisenberg's uncertainty principle states:

$$\Delta x \Delta p \geq \hbar \qquad \Delta E \Delta t \geq \hbar, \tag{6.5}$$

Where $\hbar = h/2\pi \sim 10^{-34}$ Js. Does this have any practical effect? Suppose we are driving a car. We would have determined the location of the car with an accuracy of 2 m, the speed with an accuracy of 1 km/h; the mass of the car would be 1000 kg. We would have to know the location or momentum to 18 decimal places to measure the effect!

Hydrogen fusion works in several steps:

(i) Two protons form a deuterium nucleus, which is a hydrogen isotope (isotopes of an element have the same charge number but different neutron numbers).

$$p + p \rightarrow D. \tag{6.6}$$

Deuterium has one proton and one neutron. So in this reaction a proton is converted into a neutron!

(ii) A deuterium atom reacts with a proton to form the helium isotope ^{3}He:

$$p + D \rightarrow {}^3\text{He}. \tag{6.7}$$

(iii) Two ^{3}He atoms react with each other to form:

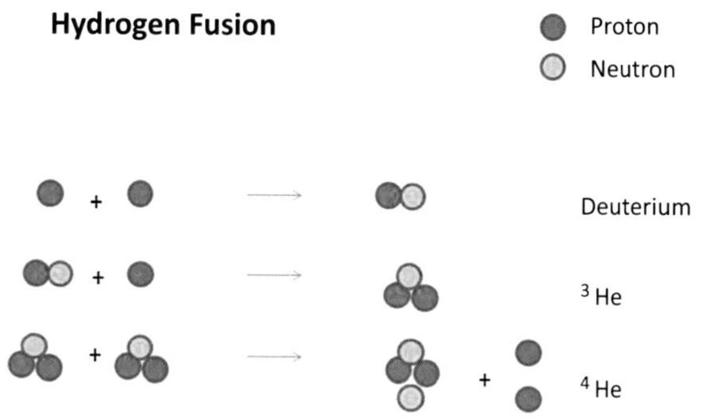

Fig. 6.13 Sketch of the three
steps in the fusion of hydrogen
to helium

$$^3He + {}^3He \rightarrow {}^4He + 2p. \tag{6.8}$$

The steps are outlined again in Fig. 6.13.

Hydrogen fusion is the energy source of the sun. Four protons form a ^4He nucleus, which consists of two protons and two neutrons. The most important thing is: The final product ^4He is slightly lighter than the sum of the initial products, i.e. four protons. The missing mass is converted into energy.

Digression The He nucleus is lighter than the four protons. What happened to this missing mass? It was converted according to Einstein's famous equation:

$$E = mc^2 \tag{6.9}$$

into energy.

If one calculates this exactly, then it follows that during hydrogen fusion about 0.7% of the mass was converted into energy. This is enough to supply the sun with energy for a total of more than nine billion years! Inside the Sun, or, as we shall see in the next chapter, inside the stars, energy is produced by the same processes as during the first three minutes when the universe was formed. Since only extremely few particles inside the sun are able to fuse due to the tunnelling effect, the sun does not explode like a hydrogen bomb, but slowly "burns" its hydrogen into helium.

Nuclear fusion is the most important process of energy production in stars. Here astrophysics has contributed significantly to nuclear fusion theory. On Earth, all energy problems would be solved if nuclear fusion could be harnessed for peaceful purposes. The problem is to keep the extremely hot plasma together long enough.

6.2.2   Solar Quakes

Why do we know relatively well about the structure of the earth? Boreholes only reach a few kilometres deep, i.e. only a few 1/1000 of the earth's radius. We know the Earth's interior relatively well from studying the propagation of earthquake waves. The propagation of these waves depends on density, temperature, and the nature of the material (whether solid or liquid). Seismology uses seismographs distributed around the world to record earthquakes.

A distinction is made between longitudinal waves and transverse waves. Longitudinal waves propagate in the direction of propagation. An example we are all very familiar with are sound waves that propagate through compressions and rarefactions of the air. In the case of transverse waves, propagation takes place at right angles to the direction of propagation, as in the case of a rope wave, for example.

Around 1960 it was discovered that there are also quakes on the sun, one speaks here of *oscillations of* the sun. Individual areas of the sun oscillate in relation to each other. These oscillations can be determined most simply by exact measurements of the Doppler effect. There are worldwide distributed observating stations to look for oscillations on the sun around the clock. The larger the oscillating area, the deeper the wave penetrates into the interior of the Sun, but these oscillations have a long period. This is sketched in Fig. 6.14. The Sun acts as a resonant body in this process. The propagating disturbances are reflected in two areas:

- Interior of the sun: Depending on the size of the oscillating area, the oscillations penetrate to different depths. In deeper layers, however, the temperature increases strongly and therefore a reflection occurs.
- Above the surface of the sun: Here the density decreases very strongly, this also leads to a reflection or the disturbances spread further upwards.

Fig. 6.14 The sun as a resonant cavity. Oscillations propagate into the interior of the sun; to different depths depending on the size of the oscillating area. They are reflected by the increase in temperature, and the wave reaches the surface again. As the density decreases greatly in the higher atmosphere of the sun, reflection also occurs. (M. Roth)

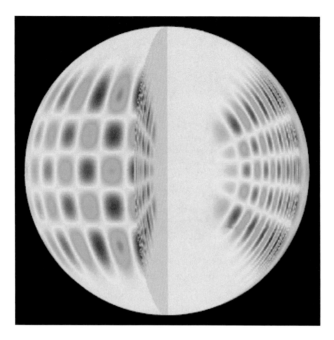

Fig. 6.15 Oscillations of the sun, so-called P-modes, and their propagation into the interior. *Red* means the area is moving away from us, *blue* means it is moving towards us. (Source: NASA)

In *helioseismology* one deals with these oscillations and can reconstruct the structure of the sun from them: One gets the values for the rotation of the sun, temperature, density, composition, pressure etc. for all layers inside the sun.

A detail on the side: We know the structure of the sun better than the structure of the earth through helioseismology. The sun is much simpler in its internal structure.

The basic period of the natural oscillation of the sun is about 5 min.

A computer-generated image shows oscillations on the Sun and their propagation into the interior (Fig. 6.15).

In Fig. 6.16 you can see how the spectral lines are shifted by the Doppler effect:

- Shift to red: when the gas (e.g. on the sun) moves away from us.
- Shift to blue: when the gas (e.g. on the sun) is moving towards us.

Today one can measure velocities of a few cm/s on the sun.

A Dopplergram of the Sun is shown in Fig. 6.17. These are velocity measurements from Doppler shifts. The data were obtained with the solar satellite SOHO (launched in 1995). The bright and dark regions indicate ascending and descending matter motions of the solar gas at the surface. The velocity due to solar rotation was subtracted.

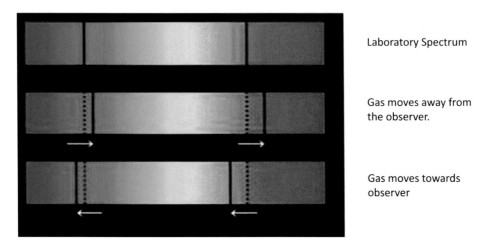

Laboratory Spectrum

Gas moves away from the observer.

Gas moves towards observer

Fig. 6.16 Doppler effect; velocities can be measured from the shift of spectral lines. Oscillations on the Sun have also been detected, which allow conclusions to be drawn about the internal structure of the Sun

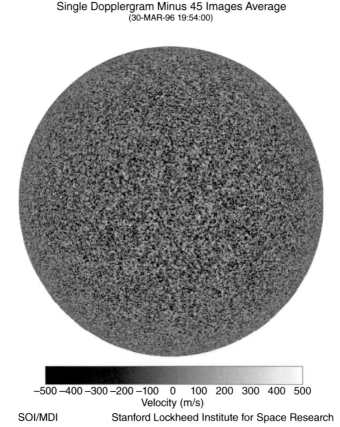

Single Dopplergram Minus 45 Images Average
(30-MAR-96 19:54:00)

−500 −400 −300 −200 −100 0 100 200 300 400 500
Velocity (m/s)

SOI/MDI   Stanford Lockheed Institute for Space Research

Fig. 6.17 Dopplergram of the Sun. The granular structure indicates different velocities at the solar surface, the magnitude of the velocities is given *below*. (Source: SOHO/MDI, NASA/ESA)

6.2.3   Neutrinos: Ghost Particles from the Sun

The following scenario was described in a science fiction film: Neutrinos, which are generated during a massive solar flare, reach the earth and heat up the earth's core. This expands and the earth's crust breaks open, with catastrophic consequences for life. Huge volcanoes form, there are floods due to tsunamis, etc. What are neutrinos actually and can something like that actually threaten the earth?

Neutrinos occur in the nuclear fusion reactions discussed earlier. When a proton is converted into a neutron, a positively charged positron (the antiparticle to the electron) is produced as well as a neutrino. This particle, as its name implies, is electrically neutral. It has very little, if any, rest mass and can pass through matter without interacting with it. Trillions of solar neutrinos pass through our bodies every second without us noticing. So neutrinos are created during nuclear fusion inside the sun and pass through solar matter as if it didn't exist. Most neutrinos also pass through the earth.

Can neutrinos be detected from the sun? This would be a possibility to observe the nuclear fusion reactions inside the sun, so to speak. One can even predict from the temperature and energy production in the interior of the Sun how many neutrinos we can expect to find on Earth. That's why experiments began 50 years ago to capture neutrinos coming from the Sun. The first experiment consisted of a water tank deep below the Earth's surface filled with perchloroethylene. A few neutrinos from the Sun cause a conversion of a few chlorine atoms to argon atoms. One has to look for these argon atoms. Another experiment measures flashes of light (Cherenkov radiation) produced by neutrinos interacting with very pure water molecules (Figs. 6.18 and 6.19). Again, it should be emphasized that of the many neutrinos that originate from the Sun, only very few cause these reactions.

The first measurements of the neutrino flux from the Sun brought a surprise: only about 1/3 of the predicted neutrino flux was observed. So are our ideas about the structure of the Sun or the nuclear fusion processes in its interior wrong? Helioseismology showed that the standard model of the Sun was correct. The solution was in the physics of neutrinos. They occur in three different states and they constantly change between these states (neutrino oscillations). At first, neutrino experiments could only measure one state (so-called electron neutrinos). The three states of neutrinos also mean that they have a finite rest mass.

Fig. 6.18 Sketch of the Japanese Kamiokande experiment for the detection of neutrinos. To ensure that the measurements are not affected by cosmic radiation, the facility is located deep underground. (Wikimedia Commons; Jnn; cc-by-sa 2.1jp)

So solar neutrinos are extremely difficult to detect and certainly cannot heat the Earth's core. They do, however, offer a great opportunity to look directly into the center of the sun, where nuclear fusion takes place.

Helioseismology and the observation of solar neutrinos provide direct insights into the interior of the sun.

Fig. 6.19 Neutrino experiment: Neutrinos captured by the Sun produce weak flashes of light that are detectable. (Kamioka Observatory, ICRR, Univ. of Tokyo)

6.3 Energy Transport

6.3.1 Energy Transport by Radiation

Near the centre of the sun, energy is generated by nuclear fusion. In addition to computer models, we have information about the interior of the sun from observations:

- Helioseismology,
- Neutrinos.

Fig. 6.20 Internal structure of the sun with core, radiation zone and convection zone. (www. szut.uni-bremen.de)

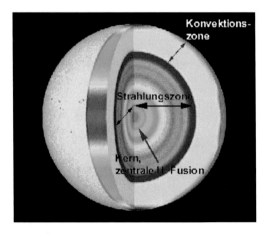

However, the energy generated during the reactions must be transported to the surface for radiation. The energy is generated in the form of extremely energetic and short-wave gamma photons. These are scattered at the atoms, re-emitted etc.. On average, such a photon travels only about 1 cm before it is scattered again. *The* zone where energy is transported by radiation is called the *radiation zone*. *Since the* photons are constantly absorbed and re-emitted, it takes about 100,000 years for them to reach the surface of the sun and appear as radiation that is partly visible to us. So we are now seeing sunlight that was actually produced inside the sun about 100,000 years ago. So, theoretically, there could be no nuclear fusion going on inside the Sun at all right now. However, by observing the neutrinos, we know that nuclear fusion is operating normally inside the Sun.

6.3.2   Convection

From a depth of just over 200,000 km below the surface of the sun, energy transport by convection begins. Hot solar plasma flows upwards to the surface, cools down, sinks back down and the process starts again. The situation is similar to boiling water, where hot bubbles form at the bottom and rise to the top. You can observe convection: The surface of the Sun is not homogeneous, but in good observing conditions you can see a cellular pattern called *granulation*. In the bright granules the matter rises upwards, in the dark intergranular spaces the matter sinks down again. The typical diameter of granulation cells is about 1000 km and they live for a few minutes.

In the transition area between the convection zone and the radiative zone is the *tachoclyne*. Strong shearing occurs in this layer. The matter above rotates like the solar surface differentially, i.e. near the equator faster than at the poles, the matter below the tachoclyne rotates like a rigid body.

The structure of the interior of the Sun is shown in Fig. 6.20. As a rule of thumb for the structure of our Sun, one can remember the following ($R_\odot$ is the Sun's radius, approx. 700,000 km):

- Core: ranges from $0 - 0.3\,R_\odot$; this is where the power generation takes place
- Radiation zone: $0.3 - 0.66\,R_\odot$; Energy transport by radiation
- Convection zone: $0.66 - 1.0\,R_\odot$; energy transport by convection.

6.4   The Surface of the Sun

6.4.1   Center to Limb Variation

First of all, how do you even define something like a surface for a ball of gas like our sun? The *photosphere of* the sun is the layer from which almost all visible radiation reaches us. This layer is only about 400 km thick, which is very small compared to the Sun's radius of nearly 700,000 km. Therefore, the limb of the Sun appears sharp to us. However, if you look closely at the solar disk (preferably using the projection method), you will see that the solar image is brighter in the center than at the limb, which is called *center to limb variation*. The explanation is simple. The temperature decreases towards the top. If you look towards the edge of the Sun, as shown in Fig. 6.21, you see into geometrically higher and therefore cooler layers that are less luminous than if you look towards the centre of the solar disk.

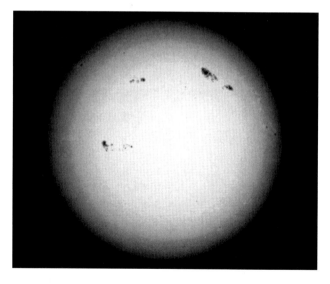

Fig. 6.21 Limb darkening of the sun. If one looks towards the limb of the sun, one sees into less deep and thus cooler layers. (Creative Commons Licence, M. Richmond)

6.4.2   Sunspots

Mostly one sees dark sunspots on the solar disk, which can become larger than the Earth. A typical image with some spots is shown in Fig. 6.21, and you can also see the limb darkening here. Spots, which reach a size of about 40,000 km, can be seen with the naked eye when the sun is low in the sky and were already known in ancient China. Spots consist of a dark core, the umbra, and a lighter filament-like penumbra. Spots appear dark because temperatures are lower there:

- Photosphere temperature: 5800 K,
- Temperature in spots, umbera: about 3800 K,
- Penumbra temperature: about 5000 K.

So if our sun would consist of only one huge spot, it would still be shining brightly in the sky, but the maximum of radiation would not be in the green but in the orange (because of Wien's law!) (Fig. 6.22).

Figure 6.23 shows a close-up of a sunspot and, in size comparison, the Earth. The dark umbra has numerous bright spots and is divided, the filamentary penumbra is clearly visible. Outside the spot, the cellular pattern of granulation can be seen.

The spots are thus up to 2000 K cooler than the surrounding solar photosphere. An explanation for this was only found when it was possible to measure magnetic fields on the Sun.

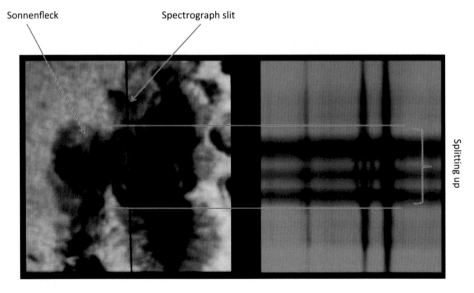

Fig. 6.22 The Zeeman effect causes spectral lines to split in the presence of magnetic fields. (www. astro.wsu.edu)

Fig. 6.23 Close-up of a sunspot. The Earth is shown at the *bottom left as* a size comparison. (Source: AAS)

Physical Concept: Zeeman Effect If you decompose the light of the sun, you can see spectral lines. These lines are created by transitions of the electrons in the atoms. In the presence of magnetic fields, these lines are split into several components. The magnitude of the line splitting depends on the strength of the magnetic field and the square of the wavelength itself. Lines in the infrared are thus more split than lines in the visible range because of their longer wavelength. This effect is called the *Zeeman effect* and can be explained by the fact that external strong magnetic fields can split the energy levels in the atoms.

Figure 6.22 illustrates the splitting of spectral lines in the region of a sunspot. On the left is the image of a sunspot, the vertical black line represents the entrance slit of a spectrograph. On the right, one sees the spectral lines split in the region of the sunspot.

Due to strong magnetic fields in the spots, the convective energy transport is hindered, so a smaller amount of hot gas reaches the surface and therefore it is cooler in the sunspots. Magnetic fields always have a north and south pole, so they are bipolar. Spots therefore occur as bipolar groups. Only rarely one observes single spots. They develop in the course of days, large spot groups even remain visible for months. For a long time it was not clear

Fig. 6.24 Differential solar
rotation. Near the surface ($r/$
$R = 1$, r ... distance from the
center, R ... solar radius) the sun
rotates faster at the equator than
at higher latitudes. From $r/R = 0$,
65 the sun rotates rigidly.
(SOHO/ESA/NASA)

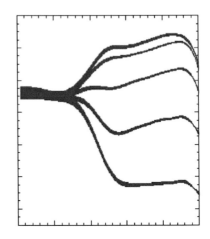

that spots are phenomena of the solar surface. Some astronomers thought they were clouds
or passing planets.

▸ The most conspicuous phenomenon on the surface of the sun: the sunspots. Because
of the strong magnetic fields prevailing in the spots, the temperature in the spots is about
2000 K lower than in the surrounding 6000 K hot photosphere.

Solar Rotation   If you observe spots for a few days, you can see how they migrate. The
rotation of the sun can be deduced from the migration of the spots. This shows that the Sun
does not rotate like a rigid body, but *differentially:* faster at the equator than at the poles.
The differential rotation of the sun was discovered by Carrington (1863) and it amounts to:

- near the equator: 24 days,
- near the pole: 31 days.

In Fig. 6.24 the course of the solar rotation with depth is given (from measurements of
the oscillations). On the *x-axis* the ratio of the distance from the centre of the sun to the
radius of the sun is plotted (r/R), on the *y-axis* the rotation frequency ν in nHz. The rotation
period in seconds is then given by $T = 1/\nu$. The values are given for heliographic latitude
0, 15, 30, 45 and 60 . Heliographic latitude is the distance from the solar equator. So
heliographic latitude 0 means solar equator. You can clearly see that the rotation period is
shorter at the equator. In the area of the tachoclyne, however, the values coincide and
further inward the sun no longer rotates differentially.

6.4.3   Granulation

In Figs. 6.23 and 6.24 the cellular granulation is clearly visible outside the spot. The name
comes from observing this phenomenon with small telescopes, where the surface of the

Sun appears granular. As discussed earlier with convection, this is convective motion. Matter rises up in the bright hotter granules and sinks down in the darker intergranules. The surface of the sun is therefore turbulent. Convection is also found in the earth's atmosphere in the troposphere.

6.4.4   Faculae

In Fig. 6.21, especially at the solar limb near spots, one also sees bright areas, the faculae. In the faculae the temperature is several 100 K higher than in the surrounding area. The best place to observe faculae is in the white light near the solar limb. Since here one looks less deeply into the atmosphere of the sun, they are therefore phenomena of the upper photosphere.

6.5   The Upper Atmosphere of the Sun

6.5.1   The Great Enigma of Solar Physics

Imagine a hot surface. Then you would expect that the further away you get from it, the temperature would decrease. Exactly the opposite is the case with the sun. First, the temperature decreases in the photospheric region and reaches the minimum of about 4500 K at an altitude of about 400 km. But then the temperature increases strongly in the two layers above:

- Chromosphere: a few 10,000 K,
- Transition region: very steep temperature increase within a short distance,
- Corona: temperature is several million K.

So there must be a mechanism that heats the upper atmosphere of the sun. This is probably a combination of so-called magnetic and acoustic heating; this will be discussed in the next chapters.

6.5.2   Chromosphere

In the past, the chromosphere could only be observed during a total solar eclipse. It appears as a reddish luminous color fringe around the sun. Its thickness is some 10,000 km and its temperature is some 10,000 K above that of the lower photosphere. One way to study chromospheric phenomena is to observe them in the *hydrogen H α line*. This spectral line is one of the most important lines in astrophysics. Its formation is explained in Fig. 6.25. When an electron in the hydrogen atom jumps from energy level 3 to 2, light is emitted at a

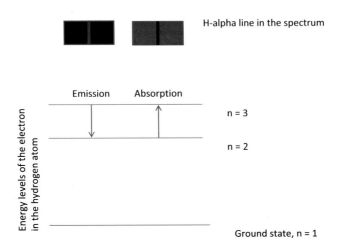

Fig. 6.25   Formation of the hydrogen line H-alpha. During the transition from $n = 3$ to $n = 2$, energy is released and a red emission line appears in the spectrum; during the transition from $n = 2$ to $n = 3$, the line appears in absorption because an amount of energy must be expended to raise the electron to this level

wavelength of 656.3 nm, which is in the red region. So objects that glow in H α are red. Due to the higher temperatures in the chromosphere of the Sun, many hydrogen atoms are in the third excited level. When the electrons then go down to $n = 2$, the H α line is produced.

With special filters one can block all wavelengths except for the light of this line and thus examine phenomena of the chromosphere on the solar disk. One sees sunspots as well as arc-shaped structures around them. These arcs are caused by matter moving along magnetic field lines. Spots are mostly bipolar groups, and the arcs represent the connection between the two magnetic poles. Due to the low plasma density in the chromosphere, matter here moves along the magnetic field lines (compare: iron filings in a magnetic field). Often in the vicinity of large groups of spots one area can be seen to light up again and again. This is called a *flare*. Flares are huge bursts of energy, mostly emitting shortwave radiation (UV, X-rays) but also charged particles. Within a few minutes, energy amounts are released that would correspond to several million Hiroshima bombs. These energies are generated by reconnection of magnetic fields. Lines of different polarities merge and thus release energy and accelerated plasma or particles (Fig. 6.26).

The formation of a flare by magnetic reconnection is shown in Fig. 6.27.

Magnetic reconnection can also be understood as a transition from a higher-energy to a lower-energy magnetic field configuration. The energy released in this process heats the solar plasma and accelerates particles.

Prominences are gas masses that appear bright at the solar limb in H alpha emission and are often arc-shaped, but the arcs can break up over the course of a few days. On the solar

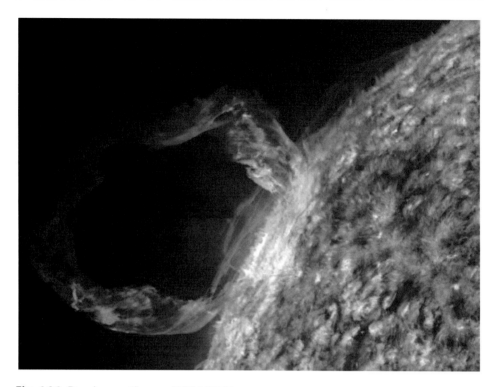

Fig. 6.26 Prominence. (Source: NASA/SDO)

disk, prominences are seen in H α light as dark filaments. Their structure is determined by magnetic fields (Fig. 6.26).

6.5.3   The Corona

The corona of the sun can be seen during a total solar eclipse as a white glowing ring of rays around the sun. The temperature increases to several million degrees within a thin transition zone, and the extent of the corona reaches several solar radii. Magnetic reconnection also occurs here and this is where the coronal mass ejections (CME) are formed. The gas density decreases from 10^{-6} g/cm (upper photosphere) to 10^{-19} g/cm^3. So the corona is actually high vacuum. What do temperatures of several million K mean then?

Almost the entire radiation of the sun comes from the 6000 K hot photosphere. Upwards the temperature in the chromosphere and corona increases strongly up to several million K.

Temperature Concept in Physics   In statistical physics, temperature is a measure of the average kinetic energy of a particle. The faster a particle moves, the higher its temperature. The kinetic (or kinetic) energy of a particle is:

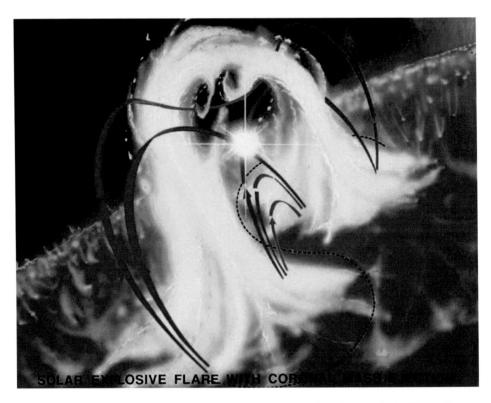

Fig. 6.27 Formation of a flare in the chromosphere by reconnection of magnetic field lines. (Source: NASA)

$$E_{\text{kin}} = \frac{1}{2}mv^2, \tag{6.10}$$

where m is the mass of the particle and v its velocity. In statistical physics, the energy of a particle is given by the formula:

$$E_{\text{therm}} = \frac{3}{2}kT, \tag{6.11}$$

where $k = 1.38 \times 10^{-23}$ J/K is the Boltzmann constant, and T *is* the temperature. This gives us the relationship between the velocity of a particle and the temperature of the gas:

$$v = \sqrt{\frac{3kT}{m}}. \tag{6.12}$$

Fig. 6.28 Course of a total solar eclipse with chromosphere *(red)* and corona, *(white,* widely extended) around the solar disk darkened by the moon. (A. Hanslmeier, Side, Turkey)

What would happen if you reached into the corona, which is several million degrees hot? Actually nothing at all, because there are only very few particles there, but then they are travelling very fast. But the transfer of momentum to the hand would be extremely small.

Figure 6.28 shows an image of the Sun during a total solar eclipse. Around the edge of the sun the red glowing chromosphere can be seen with some prominences, the white, widely extended halo is the corona.

In a CME (Coronal Mass Ejection) (Fig. 6.29), about 10^{12} g of matter is ejected at velocities of up to 3000 km/s.

At these small density values the magnetic field dominates the motion of the matter and we recognize the loop-shaped magnetic field structures in the solar corona. Because of the high temperatures, the corona can also be studied in X-ray light. We find the corona *holes,* areas that appear dark. In the corona holes (Figs. 6.29 and 6.30) the magnetic field lines are open and the fast component of the *solar wind* (which is a stream of charged particles) can escape.

Fig. 6.29 Coronal mass ejection (CME), recorded with the solar satellite SOHO. The breaking of the magnetic field arcs ejects matter into interplanetary space. (Source: SOHO, ESA/NASA)

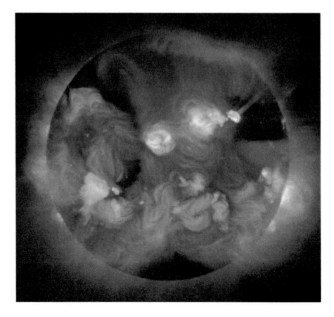

Fig. 6.30 Image of the Sun in X-ray light. You can see the hot corona of the sun, the arc-shaped structures as well as darker areas, the corona holes. (Image: YOHKOH X-ray satellite)

6.6   The Variable Sun

6.6.1   The Activity Cycle of the Sun

The activity cycle of the sun was discovered by accident. S. Schwabe wanted to find planets within the Mercury orbit about 150 years ago. Therefore he carefully observed the sun, especially sunspots disturbed him. He assumed that planets within the Mercury orbit would pass in front of the Sun (transit) and could be seen as black dots. From his records of sunspots, Schwabe eventually found the Sun's approximately eleven-year cycle of activity. Howvere this cycle is not quite regular:

- The duration of each cycle varies from less than 10 years to more than 13 years.
- The amplitudes (maxima) of the cycles are different.

As a measure of solar activity we use the *spot relative number.One* counts the number of spot groups visible on the solar disk, multiplies this value by 10 and then adds the total number of all observed spots. The value found is then corrected by a factor describing the quality of the observation, the telescope etc. . . . So if we count 3 spot groups and a total of 12 spots, the relative number is:

$$R = k(10g + f) = k(10 \times 3 + 12) = k \times 42. \tag{6.13}$$

You can see: The sunspot relative number is not really the best measure, it depends very much on whether one can still recognize faint spots at average observation conditions or whether one calculates separated spots as independent spot groups or as a single group. For historical reasons, however, this counting is adhered to and the averaged sunspot relative number is determined from observations of several observatories in order to keep the errors small.

Sunspots only occur in certain zones. At the beginning of a spot cycle they are found further away from the solar equator, at the end of a cycle they are almost at the equator. Spots are mostly bipolar groups the polarities of the preceding and following spot in terms of solar rotation are different on the northern and southern hemisphere of the Sun. If in the northern hemisphere the preceding spot is of positive polarity and the following spot is of negative polarity, then in the southern hemisphere this is the reverse. After an eleven year spot cycle this reverses, then in the northern hemisphere the preceding spot is of negative polarity and the following spot is of positive polarity. This is known as *Hale's law*. So the magnetic cycle of the sun lasts 22 years.

As can be seen in Fig. 6.31, in addition to the approximately eleven-year period of sunspots, there also appears to be an approximately 80–100-year period, the *Gleissberg cycle*. There were also times when the solar activity was extremely low and no spot was observed.

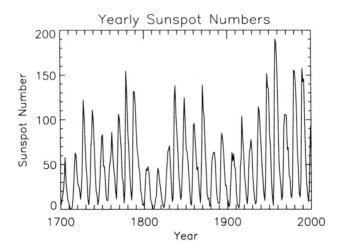

Fig. 6.31 Sunspot relative number over time. (SIDC)

The individual spot cycles are numbered consecutively. The next maximum is expected between 2024 and 2026, it is cycle 25 in the count. The previous cycle 24 started after an unusually long minimum in January 2008 and reached the maximum in February 2014. The maxima of the last cycles are getting much smaller, so it is assumed that we are approaching the minimum of a Gleissberg cycle.

So counting sunspots is the simplest measure of solar activity, but it should be emphasized that all solar activity phenomena are subject to this cycle. More sunspots mean increased occurrence of flares, CMEs, etc.

▸ Magnetic reconnection produces flares and CMEs. The Sun has an activity cycle of about eleven years.

6.6.2   Does the Sun Influence Our Weather?

Between 1645 and 1715 almost no sunspots were observed. Galileo and others observed sunspots for the first time around 1609. Then the spots fell into oblivion. The reason was that no spots were seen during the above period, which is named after its discoverer (Maunder). This period coincides with the height of the so-called Little Ice Age in the northern hemisphere of the earth. Winters were extremely cold, with reports of the Thames freezing over several times, Iceland being surrounded by ice, etc. In Europe there were cool summers and numerous crop failures. Also between 1790 and 1820 there was a long minimum of solar activity, the *Dalton minimum*. The penultimate minimum of solar activity was around 2007/2008 and lasted unusually long. Some solar physicists were already thinking of the beginning of a new grand minimum of solar activity, but eventually the Sun reached a modest maximum of activity again in 2014.

There seems to be a connection between:

Long periods of low solar activity → cooling on Earth.

Naively, one would expect the opposite: If solar activity is high, then there are many spots on the Sun, these are cooler, so the Sun should emit correspondingly less. However, this radiation deficit in the spots is overcompensated by increased radiation from the Sun in other atmospheric layers and other wavelength ranges. Thus, although the Sun has more spots during an activity maximum, it is hotter. However, while individual cycles of activity have little effect on the Earth's climate, long periods of reduced or increased solar activity can very well affect ours.

The changes of the solar radiation from minimum to maximum amount to a few per mille in the visible range, but they are much stronger in the short-wave UV and X-ray range. In addition, the strength of the Sun's *heliosphere* changes, which is the area of the solar system where the influence of the Sun through its magnetic field and solar wind predominates over other influences. With the heliosphere enveloping the entire planetary system, fewer cosmic ray particles enter the solar system. These particles could stimulate the formation of clouds in the Earth's atmosphere. However, a really precise investigation of these complex processes and interactions is still pending.

6.6.3 Solar Wind and Heliosphere

The sun emits a stream of charged particles, the solar wind. This consists mainly of protons and electrons. More than 50 years ago it was already suspected that there could be such a plasma stream from the sun: Comet tails always point away from the Sun, and one of the reasons for this feature is the solar wind. Moreover, irregularities in comet tails were found to coincide with solar activity.

Every second about 10^{36} particles are ejected from the sun, the current mass loss the solar wind is about

$$2 - 3 \times 10^{-14} \, M_\odot / \text{Jahr},$$

that's 4–6 billion tons per hour. A distinction is made between two components:

- The slow solar wind has a speed of about 400 km/s, it comes from a belt on both sides of the solar equator (streamer belt).
- The fast solar wind is about 700 km/s and it comes from the coronal holes.

The solar wind and the sun's magnetic field create a bubble in space around the solar system, the *heliosphere*. It envelops the entire planetary system and protects us from very energetic particles of cosmic radiation. When such energetic particles hit the Earth's high atmosphere, they produce the carbon isotope ^{14}C. If the heliosphere is stronger when the

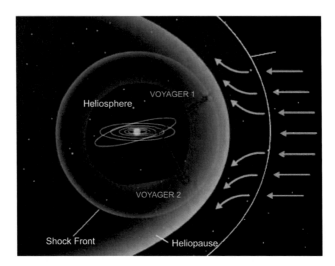

Fig. 6.32 The heliosphere encloses the entire planetary system (the outermost orbit is that of Neptune; Pluto's orbit lies a little further out and is inclined to Neptune's orbit). (NASA)

sun is very active, fewer cosmic ray particles get through and less[14]C is produced. Thus, one can reconstruct past solar activity from measurements of the carbon isotope. In Fig. 6.32 the heliosphere is sketched. The two Voyager space probes that had been studying the planets Jupiter, Saturn, Uranus, and Neptune are currently in the outer reaches of the heliosphere. The space probes were launched in 1977.

6.6.4   Sun and Space Weather

Solar activity influences the Earth's atmosphere and magnetosphere (the area dominated by the Earth's magnetic field), i.e. the Earth and the physics of near-Earth space. These influences are summarized under the term "space weather". During strong eruptions, short-wave radiation (UV and X-rays) and charged particles are emitted from the Sun. The phenomena on the Sun that are relevant for space weather:

- Flares: several per day when solar activity is high; as one flare per week near minimum solar activity. A flare produces about 2 times the energy released by the impact of comet Shoemaker Levy on Jupiter.
- CMEs (coronal mass ejections): during strong solar activity, near maximum up to three CMEs per day; near minimum activity about one CME every five days.
- Solar wind: consists mainly of electrons and protons; produces the heliosphere;

$\rightarrow$ We are protected on earth from the effects of the sun by:

Fig. 6.33 Auroras occur when charged particles from the Sun pass through the Earth's magnetic field. (After F. Olsen, Norway)

- Atmosphere: protects us from short-wave radiation; e.g. in the ozone layer the UV radiation of the sun is absorbed, therefore the temperature increases there.
- Magnetosphere: The magnetic field protects us from charged particles; some can enter the high atmosphere near the magnetic poles and cause *auroras* as shown in Fig. 6.33. There are many charged particles in the two radiation belts around the Earth *(Van Allen belts)*.
- Heliosphere: protects us from energetic particles of cosmic radiation.

During strong solar activity, satellites can become electrostatically charged, which can lead to short circuits in the electronics. The upper earth atmosphere can also heat up and expand; satellites close to the earth are thus increasingly slowed down and unstable and there is a risk of crashing or burning up in the earth's atmosphere.

The radiation and charged particles pose a threat to astronauts (e.g. in the ISS space station). These stresses will be even greater during flights to the Moon and the planned manned mission to Mars. There is no protective magnetic field on the surface of Mars. But we on Earth are also totally shielded from such influences. The ionization state of the high earth atmosphere (ionosphere) changes, it comes to disturbances in the radio communication. In order to send radio waves over long distances, it is necessary that they are reflected in higher layers of the earth's atmosphere. The reflection depends on the density of the electrons, i.e. on how many atoms are ionized in the layer concerned. This changes with solar activity and, of course, with the rhythm of day and night (Fig. 6.34). The accuracy of

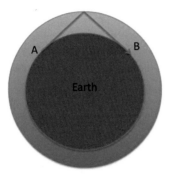

Fig. 6.34 The height of the reflection of the radio waves emitted at (**a**) determines the distance up to which they can still be received on the Earth's surface. If the ionosphere is calm, they can be received at location (**b**), if the ionosphere is disturbed (day side of the earth or strong solar storms) at location (**c**), so the range is shorter

GPS measurements is also affected. Induced overvoltages in power supply lines can cause large-scale failures in the power supply; this happened in 1978 after a strong flare outburst on the Sun: the power supply of Quebec collapsed.

What are the warning times before such events? The goal is, of course, to make a prediction about the future development of solar activity, but this is currently impossible due to a lack of observational data; moreover, the physics of the phenomena is not yet fully understood. Radiation takes only about eight minutes to travel from the Sun to Earth, so the warning time is extremely short and it is impossible to predict events on the Sun with such accuracy. Particles take up to a few days to get to Earth, depending on their speed. Here the warning time is sufficient.

▶ Flares, CMEs and solar wind are the main sources of space weather. On Earth we are protected from the influences of the Sun by the magnetic field and the Earth's atmosphere.

Stars: Formation, Structure and Evolution

7

In this section we will cover the formation, structure as well as the evolution of stars. We will see that stars still form today, that they evolve, the evolution depending on their mass. When the nuclear fuel supply is depleted, stars reach one of three possible final stages, depending on their mass: white dwarfs, neutron stars, or black holes. We will also detail the evolution of our Sun.

At the end of the chapter you can discuss:

- How long do stars live?
- How do stars form?
- Is the sun becoming a black hole?
- What are white dwarfs and neutron stars?

7.1 What Is a Star?

7.1.1 Stars: Brown Dwarfs: Planets

In the previous chapter we described in detail the star closest to us: our Sun. Table 7.1 gives a comparison of the Sun with the Earth and Jupiter. The Sun is about 10 times the size of the largest planet in the solar system and contains about 1,000 times the mass.

difference planet – star:

- Stars are much bigger than planets.
- But the most important difference is that planets do not shine themselves, without the sun the earth would be a cold, completely dark planet.

© The Author(s), under exclusive license to Springer-Verlag GmbH, DE, part of
Springer Nature 2023
A. Hanslmeier, *Fascination Astronomy*,
https://doi.org/10.1007/978-3-662-66020-1_7

Table 7.1 Comparison Sun – Earth – Jupiter

	Sun	Earth	Jupiter
Size	109 earth radii	1 earth radius	10 earth radii
Mass	333,000 earth masses	1 earth mass	330 earth masses
Temperature (surface)	6,000 K	290 K	150 K
Power generation	Hydrogen fusion	Extremely little	Little
Lights itself	Yes	No	No

Besides the distinction star – planet there is an intermediate class, the *brown dwarfs*. These are objects that are larger than normal planets, but too small to shine permanently as stars; the mass of brown dwarfs is below about 1/10 solar mass. So the hydrogen fusion inside them only ignited for a few million years. Brown dwarfs have masses of about 15–75 Jupiter masses. From about 13 Jupiter masses a deuterium fusion takes place, from 65 Jupiter masses a lithium fusion.

Digression For normal stars, you simply get their mass from the product of the volume $V = \frac{4}{3}\pi R^3$ and their density. The radius follows from:

$$M = \frac{4\pi}{3}R^3\rho \qquad R \sim M^{1/3}. \tag{7.1}$$

The greater their mass, the larger the radius. The physics of brown dwarfs is complicated. They consist of degenerate matter, so the radius is given by:

$$R \sim M^{-1/3}. \tag{7.2}$$

Brown dwarfs: The more massive, the smaller they are.
Brown dwarfs are about the size of Jupiter.

Figure. 7.1 shows a size and temperature comparison of Sun – red dwarfs, brown dwarfs and Jupiter. *Red dwarfs* are low-mass stars that glow faintly reddish, but where nuclear fusion permanently provides the energy. Since brown dwarfs have temperatures below 1,000 K, they are very difficult to find:

- They glow weakly in all wavelength ranges,
- the maximum of their radiation lies in the infrared.

▶ Stars can only exist above about 1/10 solar mass. In brown dwarfs there is nuclear fusion only for a short time.

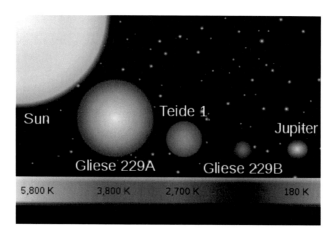

Fig. 7.1 Size and temperature comparison Sun – red dwarf, (Gliese 229A) brown dwarfs, Jupiter. (NASA)

7.1.2 Physical Properties of Stars

In this chapter we discuss some important physical properties of stars and their determination. Once again, it should be emphasized that astrophysics relies on passive observation; we can only draw conclusions about the physics of stars by analyzing their light, that is, their radiation.

Distance The distance of a star is not a physical parameter for the star itself, but it is certainly interesting and important for us if we want to imagine, for example, the structure of the Milky Way, the galaxy in which our solar system is one of many other systems. The simplest and most direct method of determining the distance of a star is the *annual parallax*. This has already been discussed. The star 61 Cygni (Cygnus is the constellation Swan) shows a relatively large proper motion in the sky. Therefore, astronomers suspected that this star might be relatively close and attempted to determine its annual parallax. This was done in 1837/1838 by the astronomer Friedrich Wilhelm Bessel (1784–1846) at the Königsberg Observatory: The parallax π is 0.3 arc seconds. This gives the distance d in parsecs to:

$$d = 1/\pi'' = 1/0.3 = 3.33\,\text{pc}. \tag{7.3}$$

A parsec (pc) is the distance a star would be if its annual parallax were 1 arcsec. A parsec = 3.36 Lj. So 61 Cygni is a good 10 light years away from us and, as we now know, it is one of the 20 nearest stars (see also Fig. 7.2).

Cygnus = Schwan

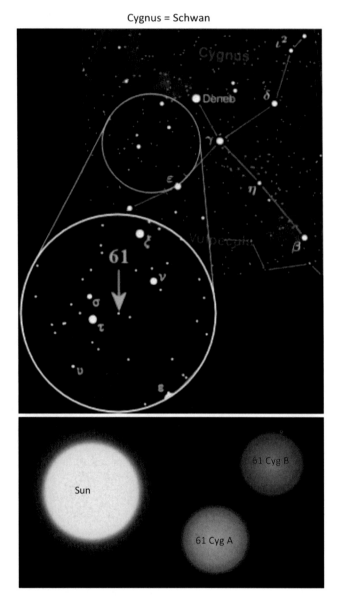

Fig. 7.2 The star 61 Cyg in the constellation Cygnus (Swan); it is a double star. (Wikimedia Commons; R. J. Hall; cc-by-sa 3.0)

Stellar Diameter If one succeeds in determining the apparent diameter of a star, then one can derive the true diameter if the distance is known. Apparent diameters of stars are very difficult to determine, since stars appear almost as points even in large telescopes. Star occultations by the moon provide a simple method. The moon moves on hour by hour about its own diameter, repeatedly occulting stars. Since stars are not exactly point-like, it takes a very short period of time (a few msec) before they disappear behind the lunar limb. Other methods use so-called interferometers.

Mass The mass of a star determines its overall evolution and also the age it can reach before it evolves into one of its three final stages: white dwarf, neutron star or black hole. Stellar masses can always be determined if the star has a companion (Kepler's Third Law).

Digression If one knows the orbital period of the companion T as well as its distance a from the star, then the mass of the star follows (M_*), if one can neglect the mass of the companion ($M_{\text{Begl}} = 0$): replaece Begl. By Comp.

$$\frac{a^3}{T^2} \sim \left(M_* = M_{\text{Comp}}\right). \tag{7.4}$$

How Hot Are Stars? We have already discussed the determination of the solar temperature in the previous chapter. Observe bright stars in the sky. Then you will notice: There are bright stars that are white to slightly bluish in color, there are yellow stars, and there are stars that are more reddish in color. We have seen that the color of a star is a measure of its temperature. White stars are therefore hotter than yellow ones, and yellow ones are hotter than reddish stars.

Digression So one decomposes the light of the stars with a spectrograph, measures at each wavelength the intensity of the radiation and from the maximum of the obtained radiation curve at a wavelength λ_{max} follows from Wien's law:

$$T\lambda_{\text{max}} = \text{const} \tag{7.5}$$

the temperature.

▸ The mass is the most important property of a star, on which its lifetime and final stage (white dwarf, neutron star or black hole) depends. Stellar temperatures are determined from the star colors.

7.2   The Brightness of the Stars

7.2.1   Apparent Brightness

Let us look at the starry sky. We see bright and fainter stars. The brightness of a star depends on

- of its truer brightness,
- its distance to us.

In Fig. 7.2 the star Deneb is drawn in the constellation Cygnus (Swan). This star shines much fainter than the brightest star of the northern starry sky Vega. Deneb is about 1,500 light years away from us, Vega only 27 light years. If Deneb would be as close to us as Vega, it would shine as brightly as the crescent moon, so it would also be visible during the day.

In ancient times, the brightness of stars was divided into *magnitude classes*. Important: This has nothing to do with the actual size of a star. The brightest stars were called first magnitude stars, then second magnitude stars, and the faintest stars, just visible to the eye, were called sixth magnitude stars. Today, this scale has been expanded to include the brightest planets as well as the Sun and Moon. Vega has the brightness 0, Jupiter at opposition about -2.5, Venus -4.5, full moon about -12 Sun -26. One subdivides the brightnesses still decimally and writes then e.g. 1^m, 0.

7.2.2 Absolute Brightness

The *apparent* magnitude is denoted by the letter *m,* which stands for *magnitudo* (Latin for size). The apparent magnitude depends on the true luminosity of a star as well as on its distance. Therefore it is not a good measure of the true luminosity of a star. The *absolute magnitude of* a star is the magnitude that a star would have at a distance of $10 \, pc = 32.6$ light years. From the difference between apparent *(m)* and absolute magnitude *(M)* one can calculate the distance *r:*

$$m - M = 2.5 \log r - 5. \tag{7.6}$$

As soon as the absolute brightness of an object is known, its distance can be determined by comparison with the apparent brightness. The absolute brightness of our sun is about 4^M, 6, so at a distance of $10pc = 32.6$ light years the sun would be an inconspicuous faint star, but still visible to the naked eye. The star Deneb mentioned earlier, on the other hand, has an absolute magnitude of -8^m, Thus Deneb would be conspicuous as a bright star even in the daytime sky and much brighter than the brightest planet, Venus. The brightest star in the sky, Sirius, has a brightness of -1^m, 4 and is about 8.3 light years away from us. The absolute brightness is then $+1^M$, 4 magnitudes.

The nearest star α Centauri is about 4.3 light years away from us. Our Earth could only be seen at this distance with the very largest telescopes, it would be an extremely faint star next to the Sun of about 24th magnitude.

7.3   Spectral Classes

7.3.1   Classification of Stars

The principle for generating a spectrum is shown in Fig. 7.3. The light coming from the sun or a star is broken up either by a prism or grating. Blue light (short wavelength) is refracted more strongly than red light (long wavelength), hence one observes the colours of the rainbow. If the resolution is high enough, dark spectral lines originating from different chemical elements can be seen. The origin of the hydrogen line H α has already been discussed. In 1802 dark lines in the solar spectrum were observed for the first time. These lines in the solar spectrum were then rediscovered and described in 1814 by the Munich optician Joseph Fraunhofer (1787–1826).

If one now analyzes the spectra of many stars, then one can introduce spectral classes. The somewhat strange order of the letters is historically conditioned: One distinguishes the following classes: O-B-A-F-G-K-M. The O stars are the hottest, the M stars the coolest. The temperature decreases from more than 30,000 K to about 3,000 K towards the M stars. In the spectra of the O and B stars one sees only a few lines, in the A stars the hydrogen lines are strongest, in the G, K, M stars the strength of lines of, for example, iron and other real metals increases. In cool stars, molecular lines are also seen. Sequence OBAFGKM is easy to remember: "oh be a fine girl (guy) kiss me".

In the hot O stars, practically all atoms are ionized, there are hardly any transitions between discrete energy levels, and the electrons are usually not bound to the atoms. Therefore we hardly see any spectral lines. Only lines of He II occur (the II simply means ionized helium, the helium atom has lost one of its two electrons). In the A stars, the

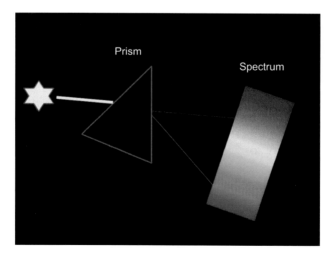

Fig. 7.3 Principle of spectral analysis. The white light of a star is broken down into the spectral colours by a glass prism (or grating). (Blue light is refracted more strongly than red)

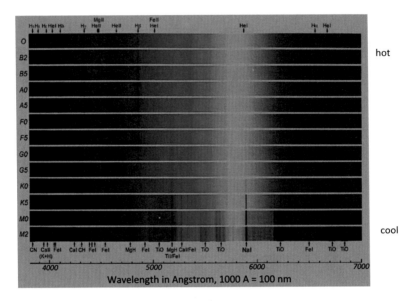

Wavelength in Angstrom, 1000 A = 100 nm

Fig. 7.4 Examples of stellar spectra. The hot O and B stars show hardly any lines. The cool K and M stars show broad molecular bands. (Wikipedia)

temperature is below 10,000 K. Here there are already hydrogen atoms with electron transitions. Because of the high temperatures many hydrogen atoms are already in the state $n = 2$ and therefore the transitions $n = 2$ to $n = 3$ (H α -line) or $n = 2$ to $n = 4$ etc. can be observed. All these lines are in the visible region and these transitions are called the *Balmer series*. For the G, K stars the temperature is below 6,000 K. Here, in addition to the Balmer lines of hydrogen, one also observes lines of other elements such as iron. Although the composition of all stars is about the same (75% hydrogen, 25% helium, less than 1% elements heavier than helium), iron lines are seen here, because such electron transitions take place because of the low temperatures. In the even cooler M stars, you also see molecular lines. Molecules can form only when the temperatures are below 4,000 K.

Figure 7.4 shows examples of stellar spectra of individual classes. Each class is subdivided again decimally. In the case of the A stars, the hydrogen line H α is most visible, while in the case of the cool K and M stars, one sees broad dark bands originating from molecules.

▸ By decomposition of the light one gets a stellar spectrum, whose dark lines allow conclusions about the temperature and composition. The spectral classes O-B-A-F-G-K-M are a sequence of decreasing temperature.

The Hertzsprung-Russell Diagram The astronomers Hertzsprung (1873–1967) and Russell (1877–1957) had the idea of plotting stars on a diagram (it was drawn up by Russell in 1913 on the basis of preliminary work by Hertzsprung): On the *x-axis* they entered the spectral type, on the *y-axis* the luminosity or absolute brightness of the stars. If

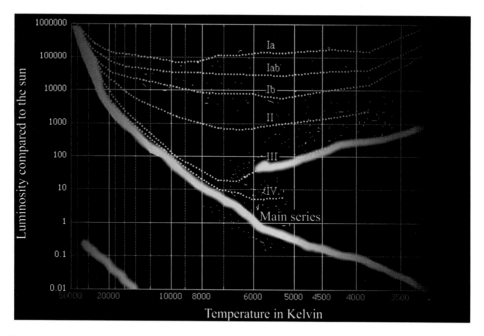

Fig. 7.5 Hertzsprung-Russell diagram (HRD)

you don't know the distance of the stars, you can also use stars of a cluster, which are all about the same distance from us. Then the distance does not matter. An analysis of this diagram yields the following:

- The majority of all stars (more than 80%) are located on a diagonal running from the upper left to the lower right. This is called the main sequence.
- Some stars are located to the right above this main sequence.
- Some stars are on the left below the main sequence.

Figure 7.5 shows the Hertzsprung-Russell diagram for many stars.

This diagram is fundamental to all astrophysics and we will explain it in more detail. Let's look at the *x-axis*, where the spectral type is plotted. O stars are on the left, M stars are on the far right. So you're dealing with a sequence of decreasing temperature. Instead of the spectral type, one could therefore also plot the temperature of stars. The temperature of a star is equivalent to its color. Therefore one could plot the so-called *color index*. This is the difference of the brightness in two wavelength ranges. On the y-axis you can plot the absolute brightness or the luminosity of the stars or the apparent brightness if all stars are equally distant.

▶ In the Hertzsprung-Russell diagram (HRD), temperature (or spectrum) is plotted against brightness. The diagonal main sequence contains the majority of stars, with supergiants and giants to the upper right and white dwarfs to the lower left.

7.3.2   Giants and Dwarfs

As shown in the HRD, there are stars that have different brightness at a given temperature. The brightness of a star (radius r) depends on two quantities:

$$L = 4\pi r^2 \sigma T^4. \tag{7.7}$$

- The term $4\pi r^2$ denotes the surface of a star. The larger the radiating surface, the brighter the star, and thus the greater its luminosity.
- The term σT^4 is the Stefan-Boltzmann law. Luminosity of a star increases with the fourth power of its temperature.

Consider two stars with the same radius. The first star has a surface temperature of 5,000 K, the second one of 10,000 K. Then the luminosity of the second star L_2 is equal to $2^4 = 16$ -times the luminosity of the first star L_1. In the HRD stars right above the main sequence stars are more luminous at the same temperature than stars on the main sequence. This can only be explained by the fact that they have a larger surface, i.e. they are much larger. That is why they are called *giant stars*. Stars that lie to the left below the main sequence are very hot, but only shine weakly, so they must be very small. These stars are called *white dwarfs*. In addition to to the classification O-B-A-F-G-K-M one defines the luminosity classes:

- I Supergiants,
- II bright giants,
- III normal giants,
- IV Subgiants,
- V Main sequence stars.

Stars are only fully defined by specifying their luminosity class.
    This fixes the position of all stars in the HRD.
    For our sun applies: spectral type G2V.
    The interpretation of the spectral type for our Sun (G2V) (see Fig. 7.6):

- G2: Hydrogen lines no longer so strong in the spectrum, our Sun belongs to the cooler stars. Lines of ionized Ca.
- V: The Sun is a main sequence star.

Fig. 7.6 Schematic of the
Hertzsprung-Russell diagram
with the approximate position of
the sun

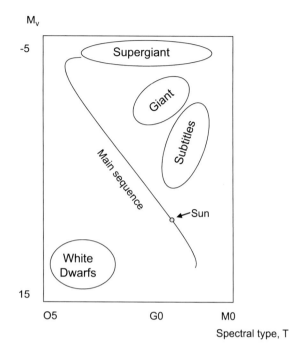

7.4   Star Evolution

Stars are very often concentrated in star clusters. A distinction is made between open star clusters and globular clusters. Open star clusters are, as the name suggests, open, irregular and contain only a few dozen stars. In the course of time they dissolve. Globular clusters are spherical, contain a few 100,000 members, and do not dissolve.

As an example, consider the open star cluster of the *Hyades,* 44 pc (1 pc = 3.26 light-years) away in the constellation Taurus. In total, more than 300 stars belong to this cluster. All stars of a cluster were formed at about the same time. The reason why you see most of the stars of this cluster on the main sequence is easily explained: the stars spend the longest phase of their evolution on the main sequence. Our Sun, a G2V star, will spend a total of about nine billion years on the main sequence. Stars hotter than our Sun use up their fuel supply much faster. Figure 7.7 shows the HRD of several open star clusters. The Hyades are about 600 million years old and thus already belong to the older open star clusters. The youngest is the open star cluster $h - \chi$ Persei, which is already visible to the naked eye on a clear moonless night, below the constellation of Cassiopeia, which appears prominently as a celestial W. The double cluster was already known to Hipparchus around 130 BC. It is about 6,000 light-years away from us. The *Pleiades* (Fig. 7.8) are also an open star cluster in the constellation Taurus and are about 100 million years old. They are about 380 light

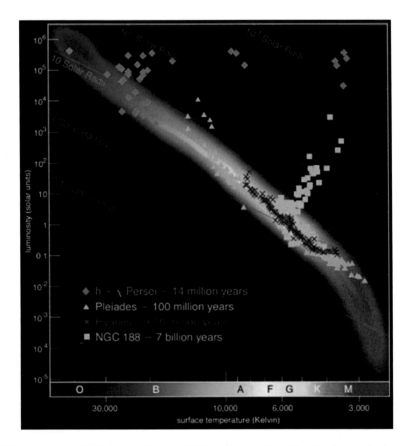

Fig. 7.7 Comparison HRD of star clusters of different ages. The older the cluster, the less massive stars are found on the main sequence. The youngest in this diagram is the cluster $h - \chi$ Per, the oldest is the cluster NGC 188. (Source: njit.edu)

years away from us. Around the young bright stars you can still see reflection nebulae, i.e. remnants of the nebula of gas and dust from which these stars were formed.

At $h - \chi$ Per the main sequence is completely occupied (violet checks). In the Hyades the main sequence branches off to the right at about spectral type A. This means: A, O and B stars have already evolved away from the main sequence. They no longer exist (more about this in the next chapter).

NGC 188 is one of the oldest known open star clusters. It is about 5,000 lightyears away from us.

As a comparison, let us look at the Hertzsprung-Russell diagram of a globular cluster. Here we see that the main sequence is no longer fully occupied and some stars are also in the region of the giants. The physical interpretation is simple: massive stars evolve faster than low-mass stars. Since globular clusters are much older than open clusters, the main sequence is no longer fully occupied up to the hottest stars.

Fig. 7.8 The Pleiades (NASA)

To summarize:

▸ The age of star clusters can be determined from the position of the branch point from the main sequence.

7.4.1   The Evolution of our Sun

The position of our Sun in the HRD hardly changes during a total of about nine billion years. It merely becomes a little brighter and moves a little further to the left along the main sequence. However, as soon as there is no more hydrogen available for nuclear fusion in the centre of the Sun, the structure of our Sun changes. Hydrogen is then fused into helium in a shell around the core. The core itself contracts, the energy released in the process heats the core, and when the temperature is high enough, helium burning begins: Three helium atoms fuse into carbon, which is

$$3\,{}^{4}\mathrm{He} \rightarrow {}^{12}\mathrm{C}. \tag{7.8}$$

Before this helium burning starts, the Sun moves slightly to the right in the HRD vertically upwards, so its temperature decreases slightly, it becomes redder, while its luminosity

Table 7.2 Main sequence stars

Mass ($M_\odot$)	Spectral type	Temperature (K)	$R/R_\odot$	$L/L_\odot$
60	O3	50,000	15	1,400,000
40	O5	40,000	12	500,000
18	B0	28,000	7	20,000
3.2	A0	10,000	2.5	80
1.7	F0	7,400	1.3	6
1.1	G0	6,000	1.05	1.2
1	G2	5,800	1	1
0.8	K0	4,900	0.85	0.4
0.5	M0	3,500	0.6	0.06
0.1	M8	2,400	0.1	0.001

increases strongly. The sun has evolved into a *red giant*. At this stage it reaches beyond the Earth's orbit, so the Earth has become part of the Sun's atmosphere and life is of course no longer possible. The age of the Sun is about 4.6 billion years, so it will take it that long again to evolve into a red giant. At the red giant stage the Sun is unstable, there are two internal sources of energy: (i) an outward-moving shell where hydrogen is burned to helium, and (ii) in the core, where helium becomes carbon. Eventually – after the shell has been repelled – almost all that remains is the core of the Sun, which becomes a white dwarf. The changes in the size of the Sun are thus:

- Sun: currently $1R_\odot$,
- Sun as red giant: about $100R_\odot$,
- Sun as a white dwarf: about $1/100R_\odot$.

The final fate of our sun is that of a white dwarf, it then gets about the size of the earth.

▶ Evolution of the Sun: main sequence star (about 9 billion years) → Red giant (a few 100 million years) → White dwarf (slowly cooling).

From the above figures we see: The Sun (like all other stars) spends most of its life on the main sequence. In Table 7.2 data for main sequence stars are given: the mass, radius and luminosity in units of the values for the Sun, as well as the temperature in K.

7.4.2   How Long Do Stars Live?

The lifetime of a star is defined as the length of time the star spends on the main sequence. As we have seen, this depends on the mass of the star.

- Massive stars burn their supply of hydrogen very quickly and live only a few million years,

- low-mass stars live for several billion years.

Digression One can estimate the lifetime of a star on the main sequence. There is a simple relationship between the luminosity of a star and its mass, which can be derived from the HRD:

$$L \sim M^{3.5}. \tag{7.9}$$

The lifetime of a star, on the other hand, is given by the ratio

$$t \sim M/L \tag{7.10}$$

and therefore one finds

$$\frac{t}{t_\odot} = \left(\frac{M}{M_\odot}\right)^{-2.5}. \tag{7.11}$$

Here $t_\odot$ means the lifetime of our sun, $M_\odot$ means the mass of our sun. Let us estimate from this formula how long a star with ten times the mass of the sun lives:

$$\frac{t}{t_\odot} = 10^{-2.5} = \frac{1}{300}. \tag{7.12}$$

Since our sun has a life span of about $t_\odot \sim$ 10 billion years, it follows for the life span of a star with ten times the mass of the sun: $t = \frac{1}{300} 10^{10} = 30 \times 10^6$ years.

A star with 10 solar masses is more than 1000 times brighter than the Sun; however, its lifetime is only 30 million years.

7.4.3   Red Giants and Supergiants

Stars with large masses evolve into supergiants. Initially, the evolution is similar to that of a normal red giant. A zone in which hydrogen fuses to helium moves outward, and the star's outer atmosphere heats up and expands. As soon as the hydrogen burning ends, a collapse occurs, the helium core heats up extremely, and helium burning to carbon starts at a temperature of about 100 million K.

7.4.4 Wolf-Rayet Stars

Massive stars with more than ten solar masses show very broad emission lines in their spectra. These were already discovered in 1867 by Ch. Wolf and G. Rayet. They are young stars with very strong stellar winds. The outer atmosphere is literally blown away. Many of these stars are components of double stars.

7.4.5 Planetary Nebulae

The name goes back to W. Herschel (around 1785), who was the first to observe these nebula-like objects with a telescope. Their appearance is often reminiscent of the small planetary discs of Uranus and Neptune that can be seen with a medium telescope, hence the misleading name. Planetary nebulae have nothing to do with planets. They are made of extremely rarefied gas that low-mass stars eject during their red-giant stage. The gas is heated by the UV radiation of the stars and therefore a nebula is seen glowing around the dying star for some time. Initially, lines of an element previously unknown on Earth were found in the spectra of these objects, which was called nebulium. Later it turned out that these are forbidden lines (e.g. oxygen and nitrogen). These lines are called forbidden because they are never observed under terrestrial laboratory conditions.

A very well-known example of a planetary nebula is the *Ring Nebula* M 57 in the constellation Lyra, shown in Fig. 7.9. It is about 2300 light years away from us. It was found in 1779 by a French astronomer and described as a large vanishing Jupiter. Its ring-shaped structure can only be seen with a telescope of at least 20 cm aperture.

7.4.6 White Dwarfs

Around 1850 it became clear that the brightest star, Sirius, must have a companion. Sirius B was first observed in 1862. It is about 10,000 times fainter than Sirius A and therefore very difficult to observe near this bright star. From the motion of Sirius B around Sirius A one could then derive its mass with the help of Kepler's Third Law: 0.98 solar masses. From the color a surface temperature of several 10,000 K results. Why does Sirius B shine so faintly? The only explanation: It has a small surface, so it is a small object about the size of the Earth. For such objects the term white dwarf has been introduced.

White dwarfs are final stages of stellar evolution for stars with a mass below 1.4 solar masses *(Chandrasekhar limit)*. The initial mass of stars evolving into white dwarfs could have been larger, but stellar winds have caused them to lose mass, leaving only a core consisting of carbon and oxygen.

It can be shown that up to this mass it is possible for so-called degenerate electrons to withstand the pressure of gravity. White dwarfs are very compact objects. A star like our sun of about 100 times the size of the earth is compressed to the size of the earth. The

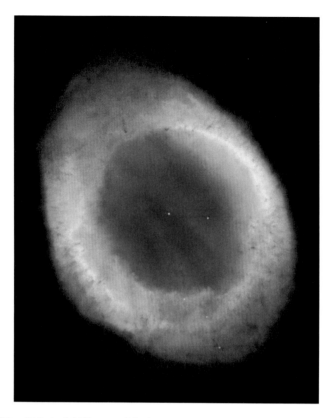

Fig. 7.9 The Ring Nebula, M 57, one of the best known planetary nebulae; the white dwarf can be seen in the centre. (NASA)

density is correspondingly high. A peculiarity of degenerate matter is that the larger the mass of white dwarfs, the smaller their diameters become. A white dwarf with 0.5 $M_{\odot}$ is about 50% larger than the Earth, a white dwarf with one solar mass has about 90% of the size of the Earth.

White dwarfs no longer produce energy, but cool down. The smaller the surface area of a white dwarf, the slower it cools. Consider a white dwarf with 0.6 solar masses.

- Brightness decreases to 0.1 $L_{\odot}$ in 20 million years,
- Brightness decreases to 0.01 $L_{\odot}$ in 300 million years,
- Brightness decreases to 0.001 $L_{\odot}$ in a billion years,
- Brightness decreases to 0.0001 $L_{\odot}$ in about 6 billion years; the star is then about as hot at the surface as the Sun. However, it is very faint and only detectable within a few pc distance. If this white dwarf were placed in place of the Sun, it would be about as bright as the full Moon.

Fig. 7.10 White dwarfs, neutron stars, black holes; formation as a function of initial mass (mass on the main sequence)

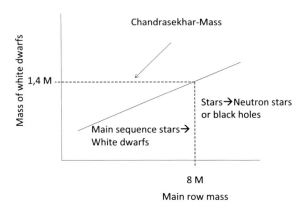

There is a lower limit of mass for white dwarfs: stars whose mass is less than 0.6 solar masses have not evolved enough since the formation of the universe to end up as white dwarfs. The more massive white dwarfs were formed from stars that originally contained up to 8 solar masses (see Fig. 7.10).

Physical Concept: Degenerate Matter Degenerate matter plays an important role in the final stages of stellar evolution, but also during other phases. Gases as we know them obey the law of ideal gases. The three state variables of a gas, pressure, volume and temperature are interrelated. Now in quantum physics there is the *Pauli exclusion principle.*

The Pauli principle states that a quantum cell can contain a maximum of two fermions which rotate in opposite directions, which is called spin. Fermions are leptons like the electron or the neutrino as well as the particles proton and neutron, which are composed of quarks.

If the density of the gas becomes very high, then this principle becomes effective. Put simply, the Pauli principle prevents, for example, a cubic meter of gas from being filled with any number of particles. From classical physics there would be no restriction, you could put any number of particles into a given volume. According to the Pauli principle, the gas begins to behave almost like a liquid. The pressure of the gas increases with density, but it is almost independent of temperature. A high density gas whose pressure is independent of temperature is called a degenerate gas. First, electrons become degenerate at a density of 10^7kg/cm^3. Neutrons become degenerate only at densities around 10^{11}kg/cm^3 . Such a density could be achieved by packing the entire earth into a cube with edge length 200 m.

Once a star is degenerate, it cools down. The internal pressure remains high, it no longer depends on the temperature.

So degenerate stars are either:

- White dwarfs: Here the electrons are degenerate,
- Neutron stars: Here the neutrons are degenerate.
- ▶ In the case of white dwarfs, the pressure of the degenerate electrons provides the counterpressure, and in the case of neutron stars, the pressure of the degenerate neutrons. Degenerate matter no longer obeys the ideal gas laws.

7.5   The Formation of Stars

7.5.1   The Solar System

Let's start with our solar system. There are some peculiarities in the planetary system that a theory of the origin of the sun and the planetary system must explain:

- The orbits of the planets lie almost in one plane.
- All planets move in the same sense around the sun; the rotation of most planets (exceptions: Uranus, Venus) is also in the sense of orbital motion.
- Planetary orbits are mostly circular.
- The Sun contains 99% of the mass of the solar system, but only about 1% of the angular momentum.
- Seen from the outside, our solar system appears to be flat.

From these facts it follows: The material from which the sun and planets were formed must also have been very flatly distributed. The angular momentum must have been transported from the sun to the outside, otherwise the angular momentum would have remained with the most massive body, the sun. So the sun rotates too slowly.

7.5.2   Molecular Clouds

There is interstellar gas between the stars. You can also find cold regions in it, where molecules can form. Since they very often occur together with dust clouds, however, the interstellar dust usually prevents the view of them in the visible range. However, many molecular clouds can be observed because of their radio emission. Besides molecules like hydrogen, H_2, carbon dioxide, CO_2, carbon monoxide, CO and even organic compounds like formic acid can be found in molecular clouds. There are very large molecular clouds, up to 10 pc in extent. These contain up to a million solar masses.

The Orion Nebula (Fig. 7.11) is the closest molecular cloud to us. It is about 450 pc away. Young hot stars are already found in it. These excite the gas clouds to glow, and we observe the Orion Nebula.

Gas clouds can be observed in different wavelength ranges:

- visible range: Mostly interstellar gas clouds glow reddish, this radiation is caused by the hydrogen line H α.
- Radio range (e.g. CO emission): Shows the distribution of the cool gas; important for star formation.
- Infrared range: distribution of the dust.

Fig. 7.11   Orion Nebula, M 42; a so-called star forming region. (NASA)

One finds in such gas clouds as the Orion Nebula cool dark regions, with temperatures of 10 K and masses between 1,000 and 10,000 solar masses. Within this there are then cooler regions, only about 0.1 pc in size and containing between 1/10 and 10 solar masses. Warmer condensations contain temperatures between 30 and 100 K, masses between 10 and 1,000 solar masses, and can be up to 3 pc in extent. In many cores of such clouds sources of intense infrared radiation are found, these are protostars. In the warmer cloud cores the stars are more massive than in the cooler ones and several stars are formed. The cloud cores collapse, and the collapse is slow because of the magnetic fields. Ionized gas (ionization occurs by nearby stars) cannot move across the field lines and friction slows down even the neutral gas.

7.5.3   Collapse of a Protostar

When a gas cloud collapses, gravitational energy is released. As long as the protostar is transparent, the generated heat is transported away as infrared radiation. Thus, practically no counterpressure is formed and the object collapses unhindered. However, the matter becomes denser and more opaque over time. Dust blocks the infrared radiation. As an example of this, consider chalk dust. If the dust is finely dispersed over a large space, then it is practically transparent, but as soon as you make the space sufficiently smaller, it becomes opaque. When the dust blocks the radiation, the protostar becomes opaque, the infrared radiation can no longer be emitted, and the pressure and temperature of the protostar begin to increase.

As soon as the pressure has become large enough to balance the gravity, i.e. the weight of the protostar, the rapid collapse of the protostar is stopped. At this stage, the protostar has only about 1% of a solar mass. But the mass is constantly increasing. Now we come to another problem. If the protostar continues to take up mass – this process is called accretion – then at some point the entire molecular cloud will have disappeared (Fig. 7.12). So we have a massive star but no planets. Will the accretion process stop? The formation of the bipolar jets is related to the magnetic field configuration and is shown in Fig. 7.13.

Fig. 7.12 Model of a young protostar with bipolar outflow along the strong magnetic field lines and an accretion disk

Fig. 7.13 Magnetic field configuration explaining the bipolar jet. (Credit: CfA)

A protostar develops strong "stellar winds". These counteract the accretion. The winds stop the accretion and blow away the remaining material around the star, the star becomes visible.

The star itself continues to contract, but no longer in free fall as before. This is called a *Kelvin-Helmholtz contraction*. This can take up to ten million years for a star like our Sun.

During the contraction phases discussed, a rotating disk (also called a *protoplanetary disk*) is formed. By contraction the disk rotates faster and faster due to conservation of angular momentum. The disk is formed by centrifugal forces. Matter falling into the protostar at the poles is not affected by the rotation of the disk. So one can easily explain the formation of a disk and thus the fact that the planets of the solar system lie almost in one plane.

7.5.4　T Tauri Stars and Stellar Winds

T Tauri stars belong to the so-called pre-main sequence stars. So these stars are not yet on the main sequence in the Hertzsprung-Russell diagram. They are cool stars of spectral types G, K, or M. Many show strong emission lines. This indicates strong activity in their chromospheres. Remember, you can observe chromospheric phenomena in our Sun outside of a total solar eclipse by emission lines formed in the chromosphere. The best known line here is the hydrogen line H α.

If one observes the brightness of the T Tauri stars, then one sees periodic changes. These are explained by huge star spots. T Tauri stars rotate faster than the sun, they need about five days. Further one observes an emission in the infrared with these stars. This points to dust, which is around these stars.

So let's take the T Tauri stars as an example of how to detect various parameters of a star from radiation, and summarize:

- Rotation: Giant starspots cause periodic changes in brightness, which can be measured;
- Chromosphere: by observing different emission lines; from their strength follows a measure of chromospheric activity.
- Dust around the star: by observing the *infrared excess;* more is emitted in the infrared than would be expected from pure radiation by the star.

Stellar winds with very high velocities are observed in young stars. The wind blows mainly along the polar axes, called a bipolar outflow, jet (Fig. 7.12). There is virtually no wind in the equatorial plane. This is important, otherwise the material to form the planets would also be blown away. Once the temperature in the core region of a star reaches about one million K, nuclear fusion begins there (essentially fusion of a deuterium nucleus with a proton). This suddenly produces high amounts of energy, which leads to strong convection of the star. Convection and rapid rotation lead to strong activity of the star, and therefore the bipolar magnetic field is very strong and ionized particles move along these magnetic field lines. Young stars lose up to $10^{-7} M_{\odot}$ per year. This is a factor of ten million higher than the current mass loss of the Sun due to the solar wind.

7.5.5   Formation of Planetary Systems in Disks

Today it is assumed that there are three processes that are decisive for the formation of planetary systems. Basically: Stars always form in molecular clouds and in groups.

- The disk evaporates due to the presence of a hot star in the vicinity. Let us assume that a disk exists around a young star in whose neighborhood a hot massive star is developing. The disk absorbs the UV radiation from the massive star, heats up, and virtually evaporates; this process takes less than 100,000 years. No planetary system is formed. In the Orion Nebula one finds many such cases.
- The star itself develops such strong winds that the disk disappears before planets could form.
- Before the disk evaporates, compressions could be formed by friction, from these first planetesimals were formed, i.e. larger lumps, km to some 10 km in size, and then the planets. Calculations show that the process of formation of the planets in our solar system took only about 10–100 million years. The age of the solar system is 4.6 billion years. The temperatures near the sun were so high that only metals and silicate-rich minerals could condense, so this is where the Earth-like planets formed; farther from the sun, other materials such as ice could form. This is the area of the gas planets with their icy moons.

Physical Concept: Gravitational Energy Gravitational energy is released as a star shrinks. This is important, as discussed, in the pre-main sequence evolution of a star. Further, this form of energy release occurs when, for example, the fuel supply of hydrogen in the core of a star has run out and the core is shrinking.

The gravitational energy of a star of mass M and radius R is:

$$E = \frac{GM^2}{R} \qquad G = 6.67 \times 10^{-11}\,\mathrm{N\,m^2/kg^2}. \tag{7.13}$$

Let us now estimate the gravitational energy of the sun:

$$E = GM^2/R = \frac{6.67 \times 10^{-1}\left(2 \times 10^{30}\right)^2}{7 \times 10^8} = 4 \times 10^{41}\,\mathrm{J}. \tag{7.14}$$

At present the sun radiates with about $4 \times 10^{26}\,\mathrm{W}$. Therefore, its energy released by contraction is sufficient for $10^{41}\,\mathrm{Ws}/4 \times 10^{26}\,\mathrm{W} = 10^{15}\,\mathrm{s}$; one year has 30 million s, so the energy is sufficient for $10^{15}/(3 \times 10^7) = 30 \times 10^6$ years. The *Kelvin-Helmholtz contraction time* for our Sun is therefore 30 million years.

7.6   Evolution of Massive Stars

In this section we discuss the evolution of massive stars. By this we mean stars whose final mass is above the Chandrasekhar limit of 1.4 solar masses. Thus, these stars do not end up as white dwarfs.

7.6.1   Nuclear Fusion in Massive Stars

Massive stars form elements up to iron by fusion. The star has a shell-like structure at the end of its evolution:

- outermost shell: consists of hydrogen and helium,
- underneath: Shell consisting of helium and carbon; temperature about 30 million K,
- underneath: Shell of carbon and oxygen,
- underneath: Shell of oxygen, neon, and magnesium; temperature about 500 million K,
- underneath: Shell of silicon and sulfur; temperature about 3 billion K,
- Iron core.

The individual elements are created by nuclear fusion. The burning time of the individual shells varies greatly. The nuclear fusion produces energy up to the element iron (Fe).

Heavier elements cannot be created by this process, or more energy is needed to create them than is gained.

Let's look at carbon burning as an example:

$$^{12}C + {}^4He \rightarrow {}^{16}O \tag{7.15}$$

$$^{12}C + {}^{12}C \rightarrow {}^{24}Mg \tag{7.16}$$

$$^{12}C + {}^{12}C \rightarrow {}^{23}Na + {}^1H. \tag{7.17}$$

We see from these reactions how carbon is turned into the other elements oxygen magnesium and sodium. Some examples of oxygen burning:

$$^{16}O + {}^4He \rightarrow {}^{20}Ne \tag{7.18}$$

$$^{16}O + {}^{16}O \rightarrow {}^{31}S + n \tag{7.19}$$

$$^{16}O + {}^{16}O \rightarrow {}^{31}P + {}^1H \tag{7.20}$$

$$^{16}O + {}^{16}O \rightarrow {}^{28}Si + {}^4He \tag{7.21}$$

$$^{16}O + {}^{16}O \rightarrow {}^{24}Mg + 2{}^4He. \tag{7.22}$$

7.6.2   A Supernova Erupts

The reactions described above lead to the formation of an iron core, which does not participate further in fusion processes. The silicon burning that leads to the formation of the Fe core takes only a few days, the hydrogen burning takes several million years even for massive stars. The iron core increases in mass, as silicon burning continues for a massive star. At first, the degenerate electrons resist the enormous weight due to gravity, and the core remains stable. However, once the iron core exceeds the Chandrasekhar mass of 1.4 solar masses, the pressure of the degenerate electrons is no longer sufficient. The nucleus implodes. At the high densities, the electrons react with the protons and become neutrons; this is called *neutronization*.

$$e^- + p \rightarrow n + \nu_e \tag{7.23}$$

e^- stands for electron, p for proton and ν_e for neutrino. In the process, the nucleus contracts very rapidly:

Fig. 7.14 Supernova 1987a, *on the left* during illumination, *on the right* an image showing only the inconspicuous progenitor. (Australian Astronomical Observatory, photograph by David Malin)

- from a few thousand kilometers to 50 kilometers within the first second,
- within a few seconds up to about 5 km.

A shock wave spreads through the entire star within a few hours. The neutrinos, which are produced during the contraction of the nucleus, escape earlier. About 1% of the total released energy is observed as a glow of the star. Except for the neutron star remaining in the core, all stellar matter is ejected by the explosion wave, and the star lights up as a type II supernova. At first the brightness of the supernova drops steeply, then slowly; the energy for this slow drop comes from the radioactive decay of the elements nickel and cobalt. A supernova thus shines much brighter than its predecessor for a few years.

On February 24, 1987, astronomers using different telescopes and at different wavelengths observed one in our Milky Way called SN 1987A (Fig. 7.14). It occurred in the Large Magellanic Cloud, which is a dwarf galaxy and part of the Milky Way. A day before the brightness burst, an increased flux of neutrinos was detected. The progenitor of the supernova was a star with about 20 solar masses, at a distance of 50,000 pc (about 150,000 light-years). The light curve of this supernova is shown in Fig. 7.15.

Fig. 7.15 The light curve of supernova 1987a. On the ordinate is plotted V, the brightness in the visual. (ESO.)

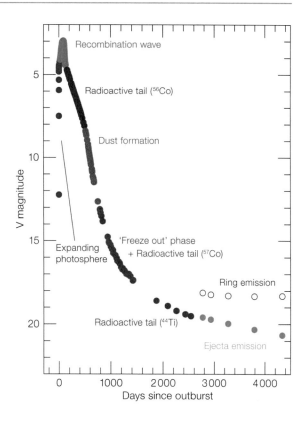

7.6.3 The Crab Nebula: A Supernova Remnant

In 1054, Chinese astronomers observed a bright star in the sky, which was also clearly visible during the day. Today we find the Crab Nebula as a remnant of this supernova explosion (Fig. 7.16). The nebula is about 6,300 light years away from us and has an extent of 11 × 7 light years. The filaments are remnants of the atmosphere of the original star and contain mostly ionized helium and hydrogen and furthermore carbon, oxygen, nitrogen, iron, neon and sulfur. The temperature of the filaments is mostly between 11,000 K and 18,000 K. The density is extremely low: about 1,300 particles per cm^3.

The filaments of the nebula expand slowly. If one compares two images taken at a great distance in time, then one can determine the expansion rate or when the explosion must have taken place.

▸ Supernovae develop into neutron stars or black holes. Only stars with more than 1.4 solar masses (in the core) become a supernova, so our sun does not!

Fig. 7.16 The Crab Nebula in the constellation Taurus is the remnant of a supernova explosion in 1054. (NASA)

7.6.4 Pulsars

In August 1967 J. Bell and A. Hewish discovered that an object in the sky emitted regular short pulses. The period of the pulses was only 0.337 s. The object was given the designation PSR 1919. At first there was great excitement, and for a short time it was even believed that these mysterious pulses were signals from extraterrestrials. Therefore the designation for this object was originally LGM-1. LGM stands for Little Green Men. The sky was searched more exactly and already within one year one found more than ten further pulsars. A little later the pulsars were identified as rapidly rotating neutron stars. Hewish was awarded the Nobel Prize for her discovery. His colleague J. Bell was left empty-handed.

Fig. 7.17 Sketch of a neutron star. The radiation is bundled by the magnetic field inclined to the rotation axis and thus results in the lighthouse effect. (According to Casey Reed)

Why do pulsars rotate in the range of seconds or even much faster? The explanation follows from the conservation of angular momentum in physics. When an iron nucleus several 1,000 km in size collapses and forms a neutron star, it must rotate very rapidly as a result of conservation of angular momentum. The situation is similar to a figure skater doing a pirouette while bending her arms to spin faster. An amplification effect due to the collapse also occurs for the magnetic field: So pulsars are stars with a very strong magnetic field that rotate very rapidly, rapidly rotating neutron stars.

But now one would expect that even if neutron stars rotate rapidly, they do not emit regular pulses. The pulses are caused by the fact that the radiation of these objects is bundled by the strong magnetic fields, i.e. originates in the area of the magnetic field lines open at the poles, and that the magnetic axis is inclined to the axis of rotation as shown in Fig. 7.17. One is dealing with a kind of *lighthouse effect*. Whenever such a beam of radiation strikes us as observers on Earth, we observe a "pulse" of radiation. The radiation itself is produced by accelerated charged particles in the magnetic field. Pulsars lose rotational energy to radiation in this way and rotate slower and slower over time. A pulsar that initially rotates with a period of 1 s (i.e., 1 Hz) rotates with a period of 2 s after

30 million years. Per rotation, its rotation slows down by 10^{-15} s. No pulsars with periods of several seconds are observed, so the pulsar phenomenon (bundled radiation in the magnetic field) only works for very rapid rotation.

▸ Pulsars are rapidly rotating neutron stars with diameters around 15 km.

Important
The term "pulsar" is misleading. Pulsars do not pulsate, i.e. they do not expand and contract; pulse in these objects refers to the radiation pulse.

7.6.5 Magnetar

Magnetars are neutron stars with extremely strong magnetic fields, the strengths of the fields exceed those of normal pulsars by several tens of thousands to millions. About 10% of all neutron stars are likely to be magnetars. There are so-called starquakes in the crust of these objects. This causes gamma ray bursts. On August 27, 1998 the magnetar SGR 1900 + 14 emitted such a gamma-ray burst. This ionized atoms in the Earth's atmosphere pointing toward the object, that is, during the night the object was in the sky. The ionization in the Earth's atmosphere reached almost daytime values, because during the day the ionization is caused by solar radiation. On December 27, 2004, the outburst of a magnetar was observed at a distance of 50,000 light-years. The gamma radiation reaching the Earth briefly had the power of the visible radiation of the full moon. In other words, within 0.1 s, the outburst produced as much energy as the Sun produces in 100,000 years. A burst of this size at a distance of 30 light-years would have caused mass extinction on Earth.

7.6.6 Type I Supernovae

Type II supernovae form at the end of the evolution of a massive star into a neutron star. Let us assume that a binary star system consists of a white dwarf and an extended companion. Matter flows from the companion to the white dwarf. As soon as the mass of the white dwarf exceeds the Chandrasekhar mass, it explodes, which is called a type I supernova. This process is sketched in Fig. 7.18.

The important point concerning both types of supernovae: stars explode at the defined central mass of the iron nucleus (or white dwarf mass) of 1.4 solar masses, because then the electrons can no longer supply the necessary pressure. This means:

Supernovae are equally bright; you can use them to determine distances.

▸ Supernovae exceed in brightness the luminosity of a whole galaxy, so they can be seen at great cosmic distances.

Fig. 7.18 Type I supernova . (According to: hera.ph1.uni-koeln.de)

7.6.7   Black Holes

Astrophysicists have been wondering for more than 100 years whether there could be stars whose gravitational pull is so strong that nothing, not even light/radiation, can leave the surface of the star. The *escape velocity from* the surface of a star with mass M and radius R is:

$$v = \sqrt{2GM/R}. \tag{7.24}$$

If we put in the values for the mass and the radius for the earth, then we find an escape velocity of 11,2 km/s. If you reach this speed with a rocket, you can fly away from the earth. If we now insert the speed of light $v = c = 300.000$ km/s for the escape velocity v, then we get for the radius R_s

$$R_s = \frac{2GM}{c^2}. \tag{7.25}$$

This is called the Schwarzschild radius. Any oject of mass M, compressed to this radius, thus appears completely "black", because not even radiation can leave the surface of the star. Let's calculate the Schwarzschild radius for our Sun. The mass of the Sun is about 2×10^{30} kg. Then we find the value 3 km for the hypothetical Schwarzschild radius of the Sun. If we compress the entire mass of the sun into a sphere of 3 km radius, then it becomes invisible, a black hole. As I said, this is purely theoretical. But if the mass of the stars at the

Fig. 7.19 Matter falls into the black hole in an accretion disk. (NASA)

end of their evolution is large enough (more than about 4 solar masses), then even the pressure of the degenerate neutrons is no longer sufficient and the star collapses into a black hole. In the center itself is a so-called mathematical *singularity;* here our concepts of space and time lose their meaning.

The question is, if nothing can escape from a black hole, how can they be detected at all? The answer is simple. Their enormous gravity causes them to bend the surrounding space. Mass can fall into the black hole, the mass does not fall directly into it, but is first in an accretion disk around the black hole (Fig. 7.19). This creates friction and produces X-rays. Black holes that are companions of other stars influence their position. Whenever strong X-ray sources are observed in the sky or a dark massive companion of more than 4–5 solar masses can be inferred from the position measurements of a star, the explanation for these phenomena is a black hole.

Black holes affect the space around them, due to strong gravity they bend it. There are three mathematical solutions:

- Black holes,
- White holes: the opposite of black holes; matter and energy flow out; exist only in theory,
- Wormholes: connect different parts of the universe with each other via so-called Einstein-Rosen bridges.

Theoretically, one could travel huge distances in the universe in a very short time through such *wormholes*, as shown in Fig. 7.20. However, this is not recommended for astronauts,

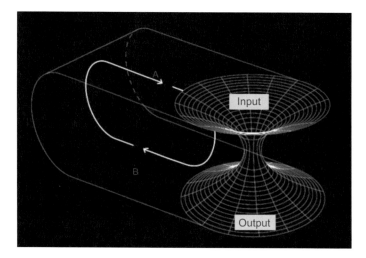

Fig. 7.20 A journey through a curved universe from A to B would be feasible by the shortest route through a wormhole

Fig. 7.21 Every mass curves space-time around itself according to Einstein's theory of general relativity. (NASA)

one would be torn apart by the ever increasing tidal forces during such a journey; physicists simply call this the spaghetti effect, i.e. the astronauts would become ever longer spaghettis.

Figure 7.21 shows that every mass curves space-time around itself. The larger the mass, the stronger the curvature. In the case of a black hole, the curvature is infinite to a singularity (Fig. 7.22).

Fig. 7.22 In a black hole, the
space-time curvature is infinite,
and there is a singularity at the
center, according to Einstein's
General Theory of Relativity

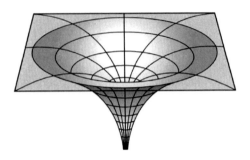

The World of Galaxies

8

In this chapter we will first discuss our cosmic home, the Milky Way. Until around 1920, it was believed that the universe consisted only of the Milky Way, and the numerous other, partly spiral-shaped nebulae were interpreted as components of our Milky Way. Only when larger telescopes (e.g. the 2.5-m Mt. Wilson telescope) were used to measure the distance of the Andromeda Nebula did it become clear that it could not belong to our Milky Way. The other galaxies were then called world islands. We will discuss different types of galaxies and their structure and formation. Galaxies arrange themselves to clusters and these again to superclusters. This then brings us to the largest structures found in the universe.

In this chapter you will learn

- what types of galaxies there are,
- how to determine distances of galaxies,
- that in quasars, giant black holes devour stars,
- what galaxy clusters are.

8.1   Our Cosmic Home: The Milky Way

8.1.1   What Is the Milky Way?

The name Milky Way comes from Greek mythology. Hera, the wife of Zeus, is said to have nursed Heracles, the illegitimate son of her husband, in his sleep. The boy behaved so clumsily that milk squirted out of her bosom and formed the Milky Way (see painting by Tintoretto (1518–1594), Fig. 8.1).

© The Author(s), under exclusive license to Springer-Verlag GmbH, DE, part of
Springer Nature 2023
A. Hanslmeier, *Fascination Astronomy*,
https://doi.org/10.1007/978-3-662-66020-1_8

Fig. 8.1 J. Tintoretto, The Origin of the Milky Way. (©Fine Art Images/Heritage Images/picture alliance)

The Milky Way can be seen in northern latitudes on clear dark moonless nights as a delicately glowing band in the sky, especially impressive on late summer or early autumn evenings. The Milky Way, which is called the Galaxy (Greek $\gamma\alpha\lambda\alpha$ means milk), was also known to other peoples. Among the ancient Germanic peoples it was called the Iring Strait, and among African peoples it was called the Backbone of Night. In 1609, the telescope was first used for astronomical observations. Galileo and others realized that the Milky Way was actually made up of a great many stars.

How is the Milky Way structured? An answer to this question tried already in 1785 W. Herschel. He simply counted stars in the sky, a method called stellar statistics. Figure 8.2 shows the picture of the Milky Way that Herschel came up with. It is wrong for two reasons:

- It was assumed that all stars had the same luminosity,
- the space between the stars is not empty, but filled with gas and dust clouds, which can strongly attenuate the light of stars located behind them.

Fig. 8.2 Herschel's conception of the Milky Way. (http://www.observadores-cometas.com/Herschel)

8.1.2 How Many Stars Are There in the Milky Way?

With the naked eye, under good observation conditions, you can see about 2500 stars on a clear night without a moon. If one resolves the Milky Way into individual stars, one has the impression that there could be an infinite number of stars.

The stars of the galaxy move around the galactic center. This lies about 30,000 light years away from us and is located in the constellation Sagittarius. The sun needs about 225 million years to orbit around the galactic center.

Digression The mass of the Milky Way can be derived from Kepler's Third Law. If M_{gal} is the mass of the part of the Milky Way that lies within the Sun's orbit around the Milky Way center, and $M_{\odot}$ is the mass of the Sun, a *is* the Sun-galactic center distance (30,000 Lj), and T *is* the Sun's orbital period around the galactic center (225 million years), then:

$$\frac{a^3}{T^2} = \frac{G}{4\pi^2}\left(M_{\mathrm{gal}} + M_{\odot}\right). \tag{8.1}$$

The mass of the sun can be neglected again.

One finds for the total mass of the galaxy about 300–400 billion solar masses. If each star contains one solar mass, this would mean the Milky Way consists of 300–400 billion stars. Most stars are low-mass, meaning they contain less than one solar mass; massive stars (O, B spectral type) are rare. First estimates of the number of stars in the Galaxy were made earlier by simply counting stars in selected sky fields.

8.1.3 The Rotation of the Milky Way and Dark Matter

The term rotation of the Milky Way means that the stars and gas clouds circle around the galactic centre. We already know such movements of celestial bodies around a center of

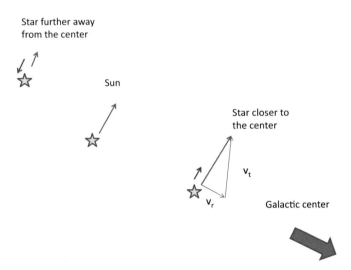

Fig. 8.3 Galactic rotation. Stars closer to the Galactic center than the Sun move faster; stars farther away move slower. Subtracting the speed of the Sun gives the red vectors

mass from our solar system: The planets move around the sun, the closer a planet is to the sun, the faster this rotation takes place.

This would also be expected of the stars in our Milky Way. Stars closer to the galactic center effectively "overtake" the Sun, and the Sun in turn overtakes stars farther away than the Sun itself. This is sketched in Fig. 8.3.

In the figure you can see the decomposition of the velocity into the two components v_r and v_t, that is the radial and the tangential component.

- Measurement of the *radial component:* This is the component of the velocity with which an object is moving directly toward or away from us; measurement by Doppler effect.
- *Tangential component:* due to displacement of the star in the sky, *proper motion.* One compares images obtained at a great distance in time and measures the position of the stars in them.

Stars that are closer to the galactic center than we are thus appear to overtake us; if they are exactly in the Sun-galactic centre line, they have no radial velocity component, as can be seen from Fig. 8.3.

Only One Fifth Visible One can now plot the rotation speed of the stars around the galactic center as a function of their distance from the galactic center. One would expect: the further away from the galactic center, the lower the rotation speed. We know this from the planetary system. Mercury moves around the Sun in only 88 days, while Jupiter takes almost 12 years to do so. Measurements of the orbital velocity of the stars showed a different picture: The rotation decreases at first, as one would expect, but at larger distances

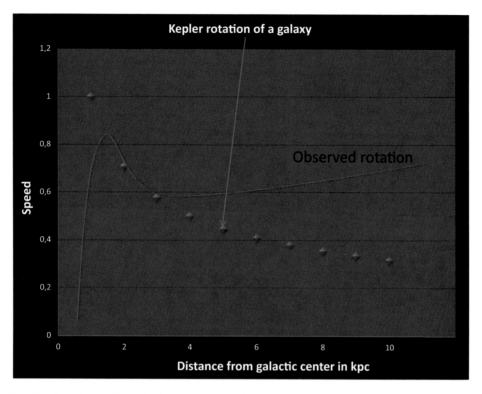

Fig. 8.4 Rotation of the Galaxy as a function of distance from the centre. The Kepler rotation is drawn in with *red dots*, the actual measured curve is shown in *red*

from the center the orbital velocity increases again. Increases in speed can only be explained by the influence of forces. Which force that makes the stars orbit faster around the center acts in the outer parts of our Milky Way?

This force apparently only acts via gravity, we cannot observe any objects there. The mass that causes this force is therefore called *dark matter.*

▸ In the outer regions of the Milky Way, there must be dark (i.e. non-luminous) matter that accelerates the orbital motion of the stars.

As we have shown, one can deduce the mass of the Galaxy from the motion of the stars. From the motion of the stars in the outer regions of the Galaxy, we can deduce that the Dark Matter is about five times the visible matter. So we can only see about 1/5 of the matter in the Galaxy, the rest is invisible Dark Matter!

Figure 8.4 shows the rotation curve of the Galaxy (red). Near the center, the Galaxy rotates like a rigid body. This can be explained by the relatively high concentration of stars there. Then follows an area where a Kepler rotation is present, but in the outer areas the rotation speed no longer decreases, but increases.

8.1.4 The Structure of the Milky Way

Viewed from the side, the Milky Way appears as a flat *disk* 40 kpc in diameter, only about 2 kpc thick. The stellar density is greatest near the galactic core. The *core* is surrounded by what is called the bulge, a spherical thickening that measures about 6 kpc in diameter. This *bulge* is elongated apart to form a bar.

The system is surrounded by the galactic *halo,* a spherical area between 30 and 40 kpc in size. In this halo we find the oldest objects of our Milky Way or the oldest objects of the universe at all, the globular clusters. As their name suggests, the stars there are arranged in a spherical cluster that is extremely stable. Globular clusters contain between a few 100,000 and a few million stars.

As mentioned earlier, our sun is about 8.5 kpc from the galactic center.

Viewed from above, the galaxy appears as a *spiral galaxy* with several spiral arms. The spiral arms consist mostly of young objects, these include:

- H-II regions: H-II means ionized hydrogen, that is, the hydrogen atom consisting of a proton in the nucleus and an electron has lost the electron due to the high temperatures. Glowing hydrogen gas is always observed in the vicinity of hot stars, whose radiation stimulates the gas to glow.
- O, B stars: As discussed in the chapter on stellar evolution, these are hot, massive stars that live only a few million years.
- Molecular clouds: Molecules can only stay in relatively cool areas, otherwise they would be destroyed by the radiation of bright hot stars. Molecular clouds always indicate places of star formation. Molecules can rotate or vibrate, and these states are quantized. The transitions are usually associated with low energies, so that molecules can be observed in the radio range.

So our sun is in the *Orion-Cygnus arm,* further in the Sagittarius arm, further out the Perseus arm. Particularly important for the study of the structure of the Milky Way is the *21-cm radiation* of hydrogen. This comes from hydrogen clouds and can be observed in the radio range at a frequency of 1440 MHz or just a wavelength of 21 cm. The origin of this radiation is interesting. One imagines that in a hydrogen atom both the proton in the nucleus rotates and the electron. Quantum physicists call this spin, so the proton has a nuclear spin and the electron has an electron spin. Now there are two possibilities (see Fig. 8.5):

1. Nuclear spin and electron spin are parallel to each other,
2. Nuclear spin and electron spin are antiparallel to each other.

During the transition from parallel to antiparallel, energy is released, namely the 21-cm radiation. The 21-cm-radiation is not absorbed by dust, therefore one can see through the absorbing dust layers and explore the structure of the galaxy. Besides gas, *dust* is also

Fig. 8.5 Formation of the
21-cm line of neutral hydrogen.
Proton and electron rotate
antiparallel to each other *(top)* or
parallel to each other *(bottom)*

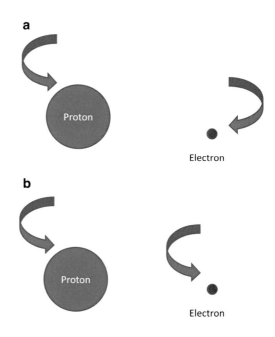

found in interstellar matter. You can easily convince yourself of this: Observe the Milky
Way in the area of the constellation Cygnus (Swan) on a clear moonless night. You can see
several dark dust clouds there, which absorb the light from the stars behind them. From the
21-cm line, one can infer the spiral structure, as shown in Fig. 8.6. If one looks from the
Earth (in our case = Sun) in the direction of the galactic centre, the individual clouds have a
different Doppler shift or intensity: the further away, the greater the Doppler shift or lower
the intensity.

▶ From the distribution of the young stars as well as the luminous hydrogen gas (H-II)
regions) and the neutral hydrogen gas (H-I regions) one can deduce the spiral structure of
our Milky Way.

8.1.5   The Monster in the Center

The center of the galaxy is located in the constellation Sagittarius. It cannot be seen in the
optical range, because dark dust clouds obscure the view. Therefore one observes the center
in the infrared or in radio wavelength ranges. These wavelength ranges are practically
unaffected by the dust. Figure 8.7 shows images taken with the 10-m Keck telescopes in
Hawaii. The position of stars near the galactic center over a period of several years is
plotted. It can be clearly seen that the orbits of the stars pass around an invisible center of
mass. Again, from the motion of the stars one can infer the mass around which they move:
it follows a mass of several million solar masses. Since we cannot see this huge mass, we
are dealing with a supermassive black *hole* (SMBH).

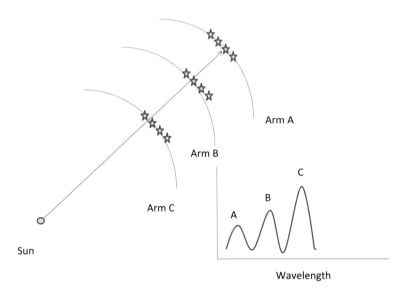

Fig. 8.6 From measurements of the profile of the 21-cm line of neutral hydrogen, one can infer different spiral arms. In this sketch, the wavelength increases from *right* to *left*

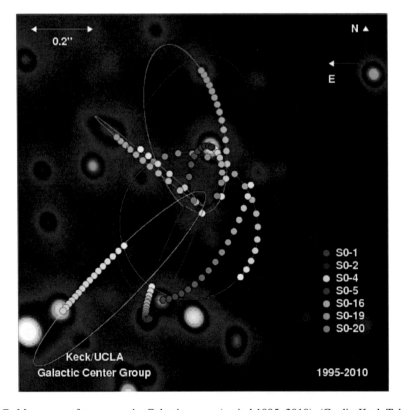

Fig. 8.7 Movement of stars near the Galactic centre (period 1995–2010). (Credit: Keck Telescope)

Does the existence of a supermassive black hole in the center of the Milky Way mean that we, along with the Sun and the other planets, will one day crash into it? The answer is no, at present this supermassive black hole hardly seems to be pulling any matter into itself, otherwise it would be visible as a source of intense X-rays. Besides, we are about 30,000 light years from the center of the Milky Way on a fairly stable orbit.

The expansion of a supermassive black hole with one million solar masses can be easily estimated from the equation for the Schwarzschild radius:

$$R_S = \frac{2GM}{c^2}.$$ (8.2)

The Schwarzschild radius for a solar mass is about 3 km. Therefore, the Schwarzschild radius for a supermassive black hole of one million solar masses is only about three million km. This is about ten times the Earth-Moon distance. At the time of its closest approach to Earth (opposition), Mars is still 20 times farther away. Such a supermassive black hole could easily fit into our solar system. However, in the course of time all planets and other objects would be sucked into it.

In the center of the galaxy one finds the radio source Sagittarius A*. The extent of this source is only 13 AU, that is 13 times the distance Earth-Sun. This is about 100 million times larger than the Schwarschild radius for a black hole with a few million solar masses. The radio emission probably comes from matter slowly collapsing into the supermassive black hole in an accretion disk around it.

The stellar density increases strongly towards the galactic center.

In a shell of 100 pc radius around that we find a stellar density of about 100 per cubic parsec. In a sphere of 10 pc radius we find several thousand stars. At a few parsecs from the center, the distance between stars decreases to light weeks.

8.1.6   The Galaxy: – A Spiral Galaxy

Seen from above, our Milky Way is a spiral galaxy with several spiral arms. The spiral arms consist of young bright luminous stars and luminous hydrogen clouds (H-II regions). Why are these objects found in the spiral arms? Luminous stars exist only for a few million years. Another problem is the apparent stability of the spiral arms. *Density wave theory* can be used to explain the structure of the arms. Density waves are something like small traffic jams on highways. A slow moving vehicle causes a traffic jam behind it. The jam moves on at the speed of the slower moving vehicle. Such a congestion point causes an increase in density. It is similar with the spiral arms of our Milky Way. In the spiral arms there is compression of matter between the stars. Stars in the arms and likewise other objects in the arms move more slowly around the center of a galaxy. So the density of matter increases in the spiral arms. Normally, one would expect the spiral arms to unwind over time. However, this does not happen because the spiral arms are always made of different material. The

Fig. 8.8 The Large and Small
Magellanic Clouds. (Creidt:
ESO/Y. Beletsky)

spiral pattern moves around the galactic center with a period of about 500 million years. So
stars form in the arms and leave them again.

Question remains, where spiral-structure resp. these density-waves come from at all.
Two explanations are offered:

- due to random disturbances,
- by close companions, satellite galaxies; the Milky Way has two close dwarf galaxies, the
  Large and Small Magellanic Clouds. However, both are only observable in the Earth's
  southern hemisphere. The Large Magellanic Cloud (LMC) is about 170,000 Lj from us
  and contains 15 billion stars; the Small Magellanic Cloud (SMC) is 200,000 Lj from us
  and contains about 5 billion stars (Fig. 8.8).

8.2 Galaxies: Building Blocks of the Universe

8.2.1 The True Nature of Nebulae

Already Christopher Wren (1632–1722) expressed in the seventeenth century the assumption that some of the observed nebulae could be gigantic star systems, which shine however
only blurred due to their large distances to us. From him (see Fig. 8.9) comes the term
world islands. He first studied mathematics, then was a teacher of astronomy and became a
co-founder of one of the most important learned societies, the Royal Society, and was also
its president from 1680–1682. He studied the buildings erected under Louis XIV in France
and then, after the great fire of London (1666), became the city's master builder. As such he
built over 60 churches and public buildings, including the new part of Hampton Court
Palace, the palace at Winchester, Kensington Palace, the library of Trinity College at
Cambridge. His major work is St Paul's Cathedral in London, built from 1675 to 1710.
He never wanted a monument to be erected to him and that is why on his tomb slab in St

Fig. 8.9 Sir Christopher Wren, mathematician, astronomer and builder of London. (©Leemage/ picture alliance)

Paul's Cathedral is the inscription *Lector, si monumentum requiris, circumspice ("Beholder, if you seek a monument, look around")*.

W. Herschel (1738–1822) is a German-English astronomer born in Hannover. In 1781 he found a diffuse object and later it turned out to be a planet beyond Saturn's orbit. In honor of the English king, Herschel named the new planet George's Star (Georgium sidus, George III was English king). This name was extremely unpopular with the French and they named the object Herschel's Star. Later they agreed on the name Uranus. Herschel observed the sky with a 6-in. telescope (1 in. $= 2.54$ cm) (see Fig. 8.10).

For his studies of non-stellar nebular objects Herschel used telescopes up to more than 1 m in diameter. He published a catalogue of 2500 nebulae and star clusters and then in 1888 the NGC, New General Catalogue, edited by J. Dreyer, appeared. This catalogue contains 7840 objects, including a great many nebulae. The largest telescope Herschel constructed had a mirror diameter of 1.26 m and a focal length of 12 m (see Fig. 8.11).

His investigations into the distribution of nebulae revealed that most nebulae are found far from the apparent position of the Milky Way band in the sky. In 1885, a star was seen shining in the Andromeda Nebula. Astronomers at the time did not believe that a single star could shine nearly as brightly as the entire Andromeda Nebula, when this nebula was a system in its own right, similar to our Milky Way, but much farther away. In 1917, astronomer Curtis observed a nova in a spiral nebula and found that the brightness curve was similar to novae in our Milky Way. However, since the nova in this spiral nebula shone

Fig. 8.10 The telescope of W. Herschel

much fainter than a nova in our system, the conclusion was obvious that they must be "world islands".

The definitive proof that galaxies exist outside our own system came from *Hubble*. He discovered Cepheids in the Andromeda Nebula. Cepheids are pulsating variable stars. They change their brightness by expanding and contracting, and in the process their temperature changes. There is an empirical relationship between their true luminosity and the pulsation period. So you can determine the true brightness of a Cepheid simply by measuring its period of brightness change. By comparing the true luminosity with the apparent measured luminosity, the distance then follows.

E. P. Hubble lived from 1889 to 1953 (USA). He studied physics and astronomy at University of Chicago, graduated with Bachelor of Science degree and then went to Oxford, England to study law. After three years he became a Master and returned to the USA. In 1904, the Mt. Wilson Observatory was founded by G. E. Hale, and from 1917, a 100-in. (2.54 m) reflecting telescope was also available there, with which Hubble made his observations of the Cepheids in the Andromeda Galaxy (Fig. 8.12).

Fig. 8.11 Herschel's largest reflecting telescope. The light is collected on the mirror at the back and can then be observed from the front through another mirror

8.2.2   Types of Galaxies

Also on E. Hubble goes back the classification of galaxies according to different types. One distinguishes:

- Spiral galaxies, these are further subdivided into
- normal spirals
- Barred spirals: In these galaxies, the nucleus is bar-shaped; the Milky Way also belongs to this type.
- Elliptical galaxies: They are clearly elliptical in shape and contain almost no dust. So you can distinguish if it is really an elliptical galaxy or just a spiral galaxy seen from the side.
- Spindle galaxies: They are spindle-shaped.
- Irregular galaxies.

Figure 8.13 shows the classification scheme. From E0 to E6 the degree of ellipticity increases more and more. Note that this is by no means an evolutionary sequence of galaxies.

Fig. 8.12 The 2.5-m Hooker telescope on Mt. Wilson (east Los Angeles), which was used to discover the expansion of the universe. (Andrew Dunn; cc-by-sa 2.0)

What is the distribution of galaxies among the different types? Here one must distinguish between the observed apparent distribution and the true distribution.

- Apparent distribution: We can still see bright galaxies at great distances, so we may have the impression that their abundance is greater.
- True distribution: One examines the distribution on different types but takes into account the distance.

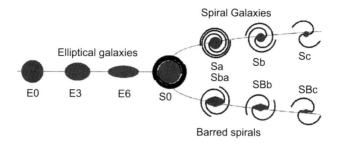

Fig. 8.13 The classification scheme of galaxies going back to E. Hubble classification scheme of galaxies

Fig. 8.14 Spiral galaxy M51 with companion galaxy. (Image: Hubble telescope)

The observed distribution shows that 77% of all galaxies belong to the spiral galaxies, but the true distribution shows only 33%. The most frequent of all types are the irregular galaxies with about 54%, but their observed frequency is only 3%, so they are rather faint and visible only at small distances. Spiral galaxies contain gas and dust, elliptical galaxies contain only very little gas and dust and mostly old stars.

Figure 8.14 shows the inner region of the spiral galaxy M51. This galaxy, which is about 30 million light-years away, is also called the *Whirlpool Galaxy*. As you can see in the

Fig. 8.15 Sombrero galaxy M104. Image: Hubble telescope

figure, it has a companion galaxy. This galaxy is very active in terms of the frequency of supernovae. A supernova was observed in each of the years 1994, 2005, and 2011. Its nucleus is relatively bright, and it is often counted among the group of *Seyfert galaxies,* which are galaxies with bright nuclei. The reddish formations visible in the image are H-II regions, i.e. regions with luminous hydrogen clouds.

Fig. 8.15 the *Sombrero galaxy.* It is about 28 million light years away from us. You can clearly see the absorption streak in the middle, which comes from interstellar matter. The extent along the major axis is 9 arcminutes, or 1/4 the diameter of the moon. It is classified as a Sa or Sb galaxy, so it is not an elliptical galaxy, but a normal edge-on spiral galaxy.

There are extensive catalogs of galaxies. The SDSS (Sloan Digital Sky Survey) contains about 930,000 galaxies. For such sky surveys, one uses special telescopes. In this case, the 2.5-m Apache Point Telescope was used. The period of the observations was 2005–2008 and the images were taken with a 150 megapixel CCD camera.

8.2.3   Why Are There Different Types of Galaxies?

The stars of our Milky Way move around the galactic center. This is observed in all spiral galaxies. In the case of elliptical galaxies, it was initially believed that they rotate very rapidly and therefore the distribution of stars is elliptically flattened. The situation would then be similar to giant planets, which rotate very rapidly and are clearly flattened. However, recent measurements show that elliptical galaxies hardly rotate at all, or the motion of the stars is almost random. In elliptical galaxies you get almost no interstellar matter and very few (if any) young stars. The reason why the content of interstellar matter is very low in elliptical galaxies is that in these galaxies the interstellar matter was consumed in an early phase of intense star formation.

Galaxies were formed by the collapse of gas clouds whose mass was about ten times that of today's galaxy. During the collapse, a random direction of rotation was established. So the scenario is very similar to star formation. However, the elliptical galaxies then lost about 90% of their angular momentum. One explanation for this would be that in the case of the elliptical galaxies, the angular momentum was transferred to the dark matter surrounding them. There may have been several clusters of stars that formed, and in the case of spiral galaxies, these stars were more evenly distributed.

8.2.4   Colliding Galaxies

Galaxies can also collide. Collisions between galaxies are relatively common. The distance of the nearest star to the Sun is 30 million times the diameter of the Sun. Therefore, collisions between stars are very unlikely. The closest companions to our Milky Way, the Large and Small Magellanic Clouds, are only about the diameter of the Milky Way away. What happens when two galaxies collide? The stars themselves do not collide, but their orbits are sometimes severely disrupted. Huge bridges of matter can form, and the formation of new stars is stimulated, known as *starburst galaxies*. Elliptical galaxies are found at the centers of large galaxy clusters, and are probably the result of such collisions. *It is also observed that large galaxies grow by "gobbling up" smaller galaxies, a kind of galaxy cannibalism. In Fig. 8.16 you can see some examples of colliding galaxies. Our Milky Way will collide with the Andromeda Galaxy within the next three billion years. It is possible that the Sun and thus the solar system will then change to the Andromeda Galaxy.

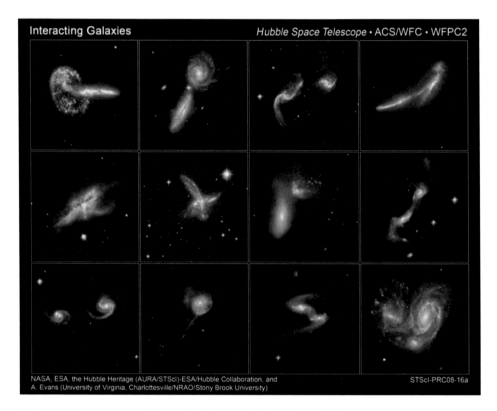

Fig. 8.16 Examples of colliding galaxies. NASA Hubble Space Telescope

8.3   Active Galaxies

One distinguishes different types of active galaxies. However, the cause for the activity of a galaxy always lies in the nucleus itself. In active galaxies, the nuclei are active.

8.3.1   Seyfert Galaxies

Around 1940, C. Seyfert observed a number of galaxies characterized by a bright, often point-like nucleus. A well-known example is the galaxy NGC 7742 (Fig. 8.17).

Seyfert galaxies emit very strong radiation in different wavelength ranges, so also in the radio range, microwaves, infrared, UV up to gamma rays. The intensity of the radiation changes within a year. It follows that the source of this radiation cannot be much larger than a light year. We are therefore dealing with a very compact object at the centre. The source

Fig. 8.17 The Seyfert galaxy NGC 7742, a galaxy with a very bright nucleus. (Source: NASA)

of the strong radiation from the nucleus is a supermassive black hole. Matter falls into the hole, friction in the accretion disk releases energy in the form of radiation.

8.3.2 Quasars

In 1940, the sky was first studied with radio telescopes. Soon some very strong radio sources were discovered. The problem with radio observations is the long wavelength of radio radiation. On the one hand, radio telescopes can be very crudely designed, since for imaging the accuracy should be in the range of $1/10\,\lambda$, that is, $1/10$ of the wavelength at which one is observing. Visible light, for example, has a wavelength of 500 nm, so the surface of the mirrors or lenses must be ground more accurately than 50 nm to ensure good imaging. Let us assume that we would investigate m-waves in the radio range, i.e. $\lambda\sim1\text{m}$, then the accuracy of the mirror of the radio antenna must only be in the cm range. However, one needs very large radio mirrors to achieve a correspondingly high resolution. It was not until around 1960 that large radio antennas became available and it was possible to precisely locate the observed radio sources in the sky. Then it was tried to observe these sources also in the optical range, and it turned out that some of these strong radio sources looked like a star in the optical range. Therefore, these objects were called quasars, from "quasi-stellar". In 1963, astronomer M. Schmidt analyzed the spectrum of quasar 3C 273, which until that time was believed to be stars. However, the analysis of the spectrum revealed a big surprise: the lines (emission lines) in the spectrum were shifted by about 0.15 to red.

Astronomers refer to redshift by the letter z, so 3C 273 has a $z = 0, 15$. In the first chapter of the book we discussed the Hubble relation: the redshift is related to the distance or escape velocity of the galaxies:

$$v = cz = dH, \tag{8.3}$$

where d is the distance and H is the Hubble constant. From the formula it is clear that the greater the redshift, the greater the distance. Therefore, the object 3C 273 must be very far away from us, it is located in the constellation Virgo (Virgo) and the distance is 2.4 billion light years. Thus it becomes clear: Quasars cannot be stars, stars can never be seen from such a distance. Meanwhile, quasars have been found with $z > 1$. Does it follow that they are moving away from us faster than the speed of light $v = c$? The answer is clearly no. At high velocities one has to calculate with the relativistic Doppler formula.

Digression Let λ_o be the observed wavelength, λ_s be the wavelength of emitted radiation at an object, then redshift:

$$z = \frac{\lambda_o - \lambda_s}{\lambda_s} \tag{8.4}$$

and

$$\frac{\lambda_O - \lambda_S}{\lambda_S} = \sqrt{\frac{1 + v/c}{1 - v/c}} - 1 \tag{8.5}$$

You can see: The redshift z can become larger than 1, but the speed at which the source moves away from us is still below the speed of light.

Why do quasars shine so brightly that they can be observed at such great distances? Quasars show brightness variations in the range of months to a few years. Therefore, the luminous source must have an extent in the range of a few light months to a few light years, i.e. it must be very compact. In the X-ray region, even shorter variations in brightness are found in the range of a few hours. Some quasars appear double in the radio range and the components appear to move at faster-than-light speeds. However, this is an optical illusion. The phenomenon of superluminal velocities in double quasars is explained in Fig. 8.18. Two quasars A and B are observed at time $t = 0$ at a distance of one billion light years. Quasar B is said to be moving towards us at a certain angle. After five years it is said to have approached us by four light-years, and in space it has travelled approximately five light-

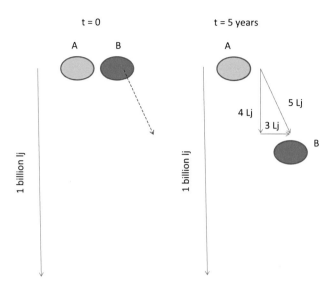

Fig. 8.18 Explanation of the superluminal velocities in a double quasar as an optical illusion

years. The distance of quasar *A* to us is still one billion light years, but that of quasar *B* is only one billion minus four years, since it has approached the Earth by four light years. So the light reaches us only one year later, after the two objects *A* and *B* have been together. However, we have the impression that *A* and *B* have moved apart by three light years during one year.

▶ The quasar phenomenon, like all other active galactic nuclei, can be explained by the assumption of a supermassive black hole at their centers.

Typical luminosities of quasars are a few hundred of those of normal galaxies, but this can go up to the value of 10,000. Black holes are very effective in terms of energy production. About 30% of the mass of the object crashing into the black hole is converted into energy. In nuclear fusion, that's less than 1%! To explain a quasar with a hundred times the brightness of a galaxy, all you need is one solar mass per year to disappear into a black hole. Since there is enough mass in a galaxy, the quasar phenomenon can be easily explained this way.

8.3.3   Gravitational Lenses

Quasars have been observed to be very close to each other and show completely identical properties: same redshift, same spectra, etc. This is explained by the *gravitational lensing effect*. When the light of a quasar passes through a galaxy cluster, it is deflected by it. The deflection of light (see Chap. 2) was one of the most important tests of the validity of relativity. This deflection of light can produce a double image or even multiple images of a quasar.

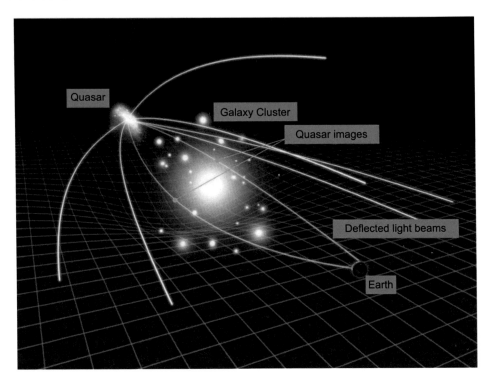

Fig. 8.19 The curvature of space of a galaxy cluster located between the more distant quasar and the Earth results in a double image of a quasar; an example of gravitational lensing

In Fig. 8.19 you can see how a galaxy cluster curves space. This causes a lensing effect when a quasar is behind it as seen from Earth, and two images of the same object are observed.

The lensing effect depends on the masses of the galaxies in the cluster. Therefore, one can infer the mass of the galaxies from the lens effect.

8.4   Galaxy Clusters

8.4.1   The Local Group

Our Milky Way, along with several dozen other galaxies, belongs to the somewhat unimaginatively named "Local Group". There are two major members in this galaxy cluster:

- Our Milky Way, Galaxy,
- Andromeda galaxy, M31.

Fig. 8.20 Galaxy M33 in the constellation Triangulum; it is the third largest galaxy in the local group. (NASA)

The Andromeda galaxy and our Milky Way contain more than 90% of the total mass of the local group.

Not far from the Andromeda Galaxy, in the constellation Triangulum, we find the smaller galaxy M33. It has a diameter of 50,000–60,000 light-years and is about three million light-years away from us (Fig. 8.20). Northeast of its center we find a star-forming region similar to our Orion Nebula (NGC 604).

All other members of the local group – altogether there should be about 60 – are much more inconspicuous. The diameter of the local group is between five and eight million light years. The Andromeda Galaxy is probably even slightly larger than our Milky Way.

Fig. 8.21 The galaxy M87 in the centre of the Virgo Cluster. The jet of matter emanating from it, more than 5000 light-years long, comes from a giant black hole at its centre. (Hubble telescope, NASA)

8.4.2   The Virgo Cluster

The Virgo Cluster is a much larger galaxy cluster in the constellation Virgo. It could contain close to 2000 galaxies and its center is about 65 million light years away from us. The objects M49, M58, M59, M60, M61, M84, M85, M86, M87, M88, M89, M90, M91, M98, M99, and M100 in the Messier catalog are galaxies of this cluster and were discovered in March 1781. At the center is the giant galaxy M87, which contains about 6 trillion solar masses (Figs. 8.21 and 8.22).

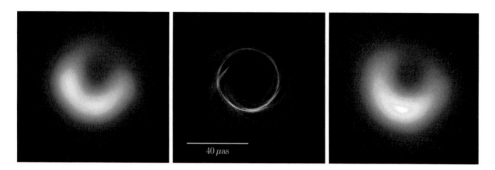

Fig. 8.22 The supermassive black hole, or more precisely the event horizon at the center of the galaxy M87. Left: Image from radio interferometer observations, centre simulation, right simulation blurred. (ESO)

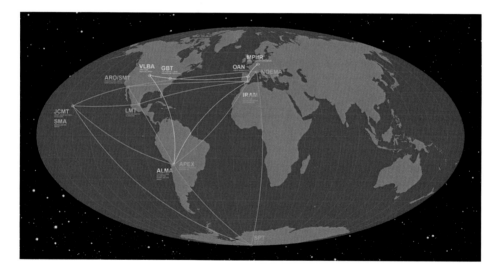

Fig. 8.23 Worldwide distribution of radio telescopes for the Event Horizon Experiment. (ESO)

The Local Group and the Virgo Cluster together form the Virgo Supercluster. So galaxy clusters arrange themselves into superclusters.

Of particular interest has been the first direct observation of a black hole in this galaxy (2017, see Fig. 8.22). Strictly speaking, this image shows the shadow of the supermassive black hole of the galaxy, which is about 55 million light-years away. It was obtained with the Event Horizon telescope (Event Horizon means event horizon). Radio telescopes were used, which were distributed over several continents. Unimaginable amounts of data had to be processed, up to 16 Gybit per second. The worldwide distribution of the telescopes is shown in Fig. 8.23.

The large distance between the telescopes makes a high resolution possible. The shadow visible in the image corresponds to the shadow of the black hole, which is about five times

Fig. 8.24 Central part of the Coma cluster. (The image was taken in infrared light with the Spitzer Space Telescope, NASA)

larger than the event horizon. The different brightness is due to the inclination of the object to the observer or to relativistic beam focusing. The ring has a diameter of 42 milliarcsec. The mass of the supermassive black hole is about 7 billion solar masses.

8.4.3   Coma Cluster

In the constellation Coma Berenice is the Coma Cluster, which contains about 1000 galaxies (Fig. 8.24). It is about 300 million light-years away from us. In the center is the galaxy NGC4889, also a giant elliptical galaxy. It is about 320 million light-years away from us.

▶ Galaxies arrange themselves into clusters and these in turn into superclusters. Our Milky Way belongs to the local group and this in turn to the Virgo supercluster.

Life in the Universe? 9

In this chapter we describe one of the most important and exciting questions of modern astronomical research: the search for life in the universe. Answering the question whether we are alone in the universe, or whether there are others, and if so, how many civilizations, would have great philosophical-ethical aspects. We will first try to answer what life is and what distinguishes it, then discuss the habitable zones as well as the possibilities of the origin of life on Earth and other celestial bodies in the solar system. Then we turn to the so-called exoplanets, planets outside our solar system. Finally, we will explain the possibilities of contact and communication.

After reading this chapter, you know,

- whether our planetary system is the exception, or planetary systems around stars are common,
- if we can make contact with aliens,
- how to find exoplanets,
- if there will be direct contact with aliens in the foreseeable future.

9.1 What Is Life?

9.1.1 Definition of Life

The question, what is life, cannot be answered in one sentence. Life is characterized by many properties:

© The Author(s), under exclusive license to Springer-Verlag GmbH, DE, part of
Springer Nature 2023
A. Hanslmeier, *Fascination Astronomy*,
https://doi.org/10.1007/978-3-662-66020-1_9

- Growth,
- Replication, reproduction,
- Metabolism,
- Reactions to stimuli.
- Life on earth is based on the construction of the smallest units, the cells. There is unicellular and multicellular life, in the latter certain cells have taken on specific properties (e.g. in humans: nerve cells, skin cells, muscle cells, etc.).

So far, we know of life on only one celestial body in the universe: Earth. This life is based on two fundamental elements:

- Water: This molecule has important properties for life. It can dissolve substances, has a large heat capacity and occurs frequently in the universe. Without water in liquid form, terrestrial life is unthinkable.
- Carbon compounds: The chemical element carbon can form very complex compounds called organic compounds. Up to four atoms can be attached. E.g. methane, CH_4, or ethane, C_2H_6, proteins, sugars, fats, etc. Organic compounds are also found very frequently in the universe: in the atmosphere of Saturn's moon Titan, on the surface of Jupiter's moon Europa, on meteorites, Mars, even in interstellar matter, interstellar molecular clouds, etc.

$\rightarrow$ The basic building blocks of life as we know it on Earth, water and carbon, are very common in the universe, life should therefore be common.

One therefore looks for these compounds or tries to check the basic conditions of life as we know it on earth, if one looks for life on other planets. Is there water in liquid form on other planets, are there signs of metabolism, reproduction, etc.? The first probes to soft land on Mars were the Viking landers. They were looking for organic compounds and water.

The Perseverance mission, which landed on Mars in February 2021, will search for life in the Jezero crater, which may once have been filled with liquid water (Fig. 9.1). Whether life exists on Mars remains controversial. In an early phase, Mars could have been covered by large expanses of water, the climate was much more favorable, and if life developed, it could have retreated into niches.

In Table 9.1 a list of solar system objects is given and whether the two basic building blocks of life, water and organic compounds, were found there is entered. It can be seen that at least the basic building blocks of life are present in these selected objects.

9.1.2 Origin of Life on Earth

Earth, like the solar system, is about 4.6 billion years old. The formation of the earth took about 500 million years and during the early phase there was a cosmic bombardment of meteoroids, asteroids comets. It is possible that we owe the water on Earth to the comet

Fig. 9.1 First images of the Perseverance mission landing in a Martian crater in February 2021

Table 9.1 Some selected objects in the solar system where water and organic compounds exist

Object	Water	Organ. Liaison.	Temperature range
Earth	Yes	Yes	Favorable, above 0 °C
Mars	Instant sublimation	Yes	Currently mostly below 0 °C
Venus	Water vapour in atmosphere	Yes	Currently too hot
Jupiter	Steam	Yes	Very cold
Europe	Ocean below ice crust	Yes	Below ice crust favorable
Titanium	Ocean under surface	Yes	Extremely low
Comets	Ice	Yes	Evaporation near the sun

impacts from this period, because when the Earth formed it was too hot and any water that was present evaporated. Our moon was formed by the collision of the earth with a protoplanet about the size of Mars. The oldest fossils found are about 3.5 billion years old. So, roughly speaking, it took almost a billion years for life to develop on our planet.

The experiment of Urey and Miller was conducted in 1952/1953 at the University of Chicago. A container filled with water, methane, ammonia and hydrogen was heated. These substances were supposed to reproduce the conditions in the early Earth's atmosphere. Electrical discharges simulated lightning. After a week, the mixture contained more

Fig. 9.2 Structures of mineral deposits formed by volcanic outgassing on the deep ocean floor. The exit temperature of the gases is 400°. (Center for Marine Environmental Sciences, Univ. Bremen)

than ten percent organic compounds, i.e. the basic building blocks of life. It was later shown that UV radiation from the sun also produces such compounds. The atmosphere of Saturn's moon *Titan* is very dense and contains tholins, which are organic compounds produced by the interaction of sunlight with the components of Titan's atmosphere.

Today, the theory of the origin of life in the *black smokers is* favoured, which are geyser-like degassings on the sea floor (see Fig. 9.2). Hot gases flow out and various compounds settle in the cold water of the seafloor, giving the impression of an escaping dark gas (hence black smokers). Many bacteria have been found to thrive in such extreme environments. These are called *extremophiles*. There are bacteria that develop excellently at high temperatures, so-called thermophiles, or in very salty environments so-called acidophiles.

The emergence of life on the ocean floors also offers the advantage that it would have been protected there from the UV radiation of the sun. The early Earth's atmosphere did not yet contain free oxygen, and so no ozone layer could develop to protect us from the sun's UV radiation. Life therefore originally arose in water.

Life arose on Earth 3.5 billion years ago. Living organisms (e.g. cyanobacteria) developed rapidly and released free oxygen into the Earth's atmosphere through photosynthesis. Slowly, the Earth's atmosphere became enriched with free oxygen. About a billion years ago, a thin ozone layer formed and the harmful short-wave UV radiation from the sun could no longer penetrate to the earth's surface.

9.1.3   Earth's Protective Shields

Life reacts very sensitively to incident high-energy short-wave radiation or to high-energy particles. The radiation damage can only in slight cases be corrected by the affected organisms themselves, reproductive errors develop or cancer occurs. We are protected on the earth's surface from these influences by:

- Atmosphere: By absorbing (e.g. UV radiation in the ozone layer or X-rays in higher layers) short-wave radiation, the atmosphere offers us protection from the radiation. Some planetary moons may have an ocean of salty water beneath an ice crust (e.g. Jupiter's moon Europa). Here the ice crust takes over the protective role of a missing atmosphere.
- Earth's magnetic field: Electrically charged particles are deflected by magnetic field lines; as a rule, they cannot penetrate these field lines.
- Heliosphere: The sphere of influence of the solar wind and the magnetic field of the sun extends over the entire planetary system and also deflects high energy particles of cosmic radiation.

In addition to shielding us from high-energy radiation, the Earth's atmosphere naturally also has a balancing effect on the global temperature of the Earth. Without the natural greenhouse effect, it would be up to 30° cooler and the Earth would be a frozen ice planet. The clouds protect against excessive cooling during the night.

9.2   Habitable Zones

9.2.1   What Is a Habitable Zone?

We assume, in the absence of other knowledge, that life is linked to the presence of water in liquid form. Then we can introduce habitable zones, areas around an object (star or giant planet) where water can exist in liquid form. Note: life also requires energy. In most cases, this energy comes from the star around which a planet orbits, which in our case is the Sun. But it could also come from heating of the celestial body due to strong tidal forces, as in the case of some moons of Jupiter or Saturn.

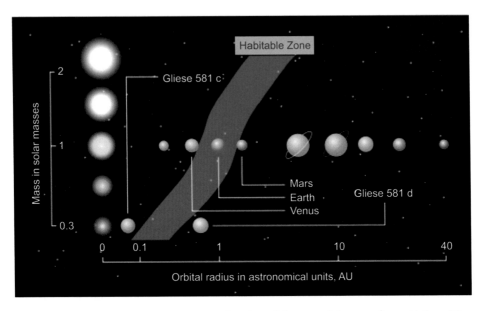

Fig. 9.3 The extent of the habitable zone as a function of the mass of the central star. (Adapted from www/astrobio.net)

9.2.2   Circumstellar Habitable Zones

Let us consider a star with a given temperature and ask the question, at what distance from this star could water exist in liquid form on a hypothetical planet. It is obvious that this habitable zone would have to be very close to the star if its temperature is relatively low. Stars that are cooler than our Sun have habitable zones that are relatively close to the star. For hot stars, the habitable zone moves outward. The cooler the star, the less extended is the habitable zone and the less likely it is to find a suitable planet there.

Figure 9.3 shows a sketch of the habitable zone. For comparison, the planets of the solar system are shown. It can be seen that Venus is too close to the Sun and Mars is just outside the zone. Furthermore, the two exoplanets Gliese 581 c and Gliese 581 d are shown in this figure.

Stars with 0.3 solar masses have a habitable zone at only 1/10 the distance Earth-Sun, or 15 million km. This is 40 times the distance Earth-Moon. Planets this close to the star are highly susceptible to changes in stellar brightness, outbursts on the star, etc. For a two solar mass star, the habitable zone moves to about the distance Jupiter is from the Sun. Note, however, that the lifetime of stars decreases with increasing mass, and thus massive stars evolve so rapidly that there is no time for life to form.

Fig. 9.4 The European
spacecraft Juice will explore the
Jupiter system for three years
starting in 2030. (ESA)

9.2.3   Circumplanetary Habitable Zones

Due to the strong tidal forces in the vicinity of a large planet, satellites of this planet can be
heated, since the moon is constantly deformed and its mass is practically kneaded. In the
case of Jupiter, this strong tidal effect is seen through volcanism on Io (sulphur volcanoes)
or the presence of a liquid ocean beneath an ice crust on the moon Europa and other moons.
Tidal power here provides the necessary energy for life. Whether life has actually evolved
in Europa's ocean will only be determined by future space missions. The planned JUICE
mission will play an important role in this. The mission is scheduled to launch in 2023 and
target Jupiter's moons starting in 2029. In 2032, it will enter orbit around Jupiter's moon
Ganymede (Fig. 9.4).

9.2.4   Galactic Habitable Zone

Our sun, and thus the solar system, is about 30,000 light years away from the center of the
Milky Way. It is assumed that there is something like a galactic habitable zone. Too close
to the center of a galaxy, conditions are unfavorable for the emergence of life. The stellar
density becomes higher and higher, and disturbances can allow objects such as those we

know from the Oort cloud enveloping the solar system to enter the interior of a planetary system and wipe out life by impact. So the probability of comet showers becomes greatly increased. Stars near the galactic center can explode, and the short-wave radiation that results is equally hostile to life.

On the other hand, if the distance from the galactic center is too large, then the formation of planetary systems is questionable. The content of elements heavier than helium decreases towards the outside of a galaxy. Without elements heavier than helium, there are no planets with solid surfaces. So a habitable zone exists in a spiral galaxy. Elliptical galaxies are hardly suitable for habitability, because they contain only a few elements heavier than helium.

▸ In the search for life, one defines habitable zones. Planets within these zones could hold water in liquid form on their surfaces. For planets to form around a star at all, there must be a certain distance from the galactic center.

9.3   How to Find Exoplanets?

In this section we deal with the search for exoplanets. It is interesting that one does not need any complex telescopes for such research, but can already detect exoplanets at least indirectly with relatively simple means.

Exoplanets can only be seen directly in exceptional cases. Mostly their faint glow is outshone by the much brighter central star, and because of the distance since exoplanets always appear very close to the central star.

9.3.1   Transit Method

We have already discussed that in rare cases we can observe the transit of the dark Venus or Mercury disk in front of the Sun. So, if our line of sight lies in the direction of the orbit of an exoplanet (by this we mean planets outside our solar system), then planetary transits may occur, which can be measured by a very small decrease in the brightness of the star. The size of the star can be determined from the duration of the eclipse, and the size of the planet from the duration of the brightness decay. The more precise the brightness measurements, the smaller planets can be found. The most accurate measurements are obtained from space. As seen from Earth, large exoplanets also show phases. Figure 9.5 shows the observation of a transit. At the top of the figure you can see the decrease in brightness due to the transit of the planet, the light of the star is dimmed as the planet passes. In the middle you can see the secondary minimum, when the planet is behind the star; then only the light of the star reaches us and at the very bottom you can even see the change in brightness due to the illumination phase of the planet. If the planet is exactly between earth and star it is not illuminated, shortly before it disappears behind the star seen from us it is fully illuminated.

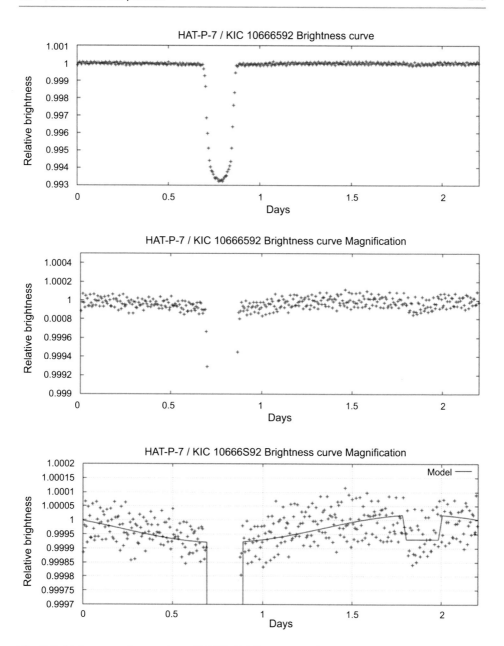

Fig. 9.5 Light curve of an exoplanet (HAT P-7). The more precise the observations, the more details become visible. (According to M. Kuhlberg. Kepler satellite data)

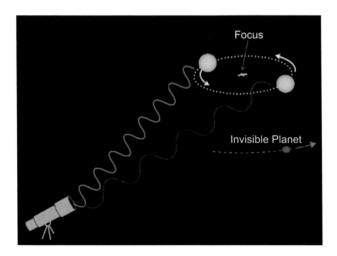

Fig. 9.6 Centre-of-mass motion of a star due to the exoplanet. By exact measurement of the radial velocity the mass of the exoplanet follows. (Adapted from ESO)

9.3.2 Radial Velocity Method

Figure 9.6 shows a star and an invisible exoplanet orbiting around it. Both move according to the laws of physics around the common center of gravity. If the star moves away from us during this motion, its spectral lines are shifted to red. If it moves towards us along its orbit, the lines are shifted to blue. By very precise measurements one can determine this movement and estimate from it also the mass of the exoplanet. If there are several periods in the motion around the center of gravity, this indicates several exoplanets in this system.

9.3.3 Stars Change their Position

This method is coupled with the radial velocity method (see Fig. 9.6). The motion of the star around the center of mass of the system causes slight changes in position on the sky, which change periodically. These effects are greatest for very large massive planets that are very close to the star. As the ratio of stellar mass to planetary mass becomes smaller, the effects become larger. Many of the exoplanets found so far are Jupiter-like objects very close to their parent star. Earth-like planets are thus difficult to find.

   The motion of the center of gravity of the sun (the sun's core) is shown in Fig. 9.7. The motion is complicated because our solar system contains eight major planets. So the centre of gravity of the solar system is constantly moving and may be just outside the sun itself.

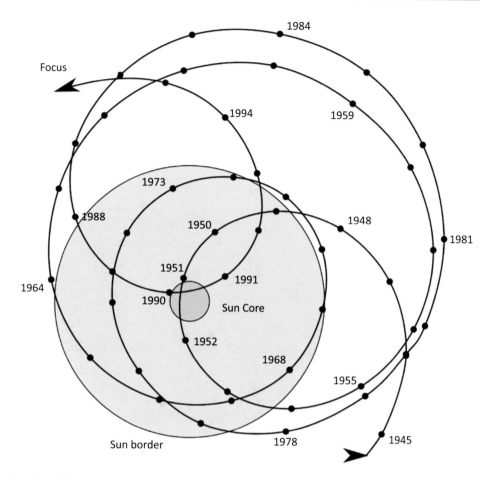

Fig. 9.7 Movement of the centre of gravity of the sun's core, caused by the movement of the planets. The *yellow large circle* represents the size of the sun. (Adapted from P. Horzempa)

9.3.4   Satellite Missions

In 2009, the KEPLER mission was launched. Over a period of more than four years, 145,000 stars were observed in a selected star field. The field was chosen to be as far away as possible from interfering objects of our solar system (e.g. small planets) and to be located in a star-rich region near the plane of the Milky Way (Fig. 9.8).

The GAIA mission (Global Astrometric Interferometer for Astrophysics, Fig. 9.9) was launched in December 2013. It is intended to measure one billion stars precisely. It is hoped to find several 10,000 exoplanets.

▶ Exoplanets are found, for example, by observing transits or from radial velocity measurements. Direct observation is extremely difficult.

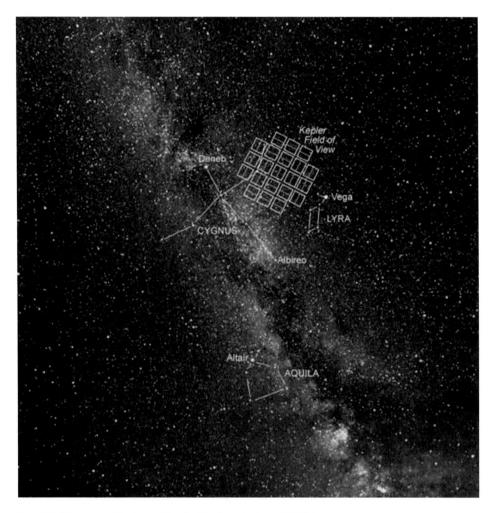

Fig. 9.8 The star field selected for the Kepler mission. (NASA)

9.3.5 How Many Exoplanets Have Been Found?

The number of discovered exoplanets is constantly increasing. For orientation, we give the status as of 31 January 2021:

- secured discoveries: 4408
- Multiple planet systems: 721
- Earth-like exoplanets: 230

Figure 9.10 shows an overview of the discovered exoplanets (mass) and the mass of the star.

Fig. 9.9 GAIA, a satellite designed to accurately measure one billion stars. (NASA)

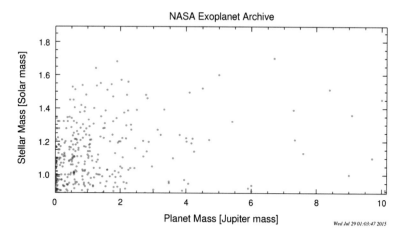

Fig. 9.10 Statistics of exoplanets. Mass in units of Jupiter mass versus stellar mass. (NASA)

9.3.6   Examples of Exoplanet Systems

The Trappist-1 System Trappist-1 is a cool star with a surface temperature of only 2500 K and 0.09 solar masses and about 1/10 of the solar radius. The star can only be observed with large telescopes, its brightness is 18.8 magnitudes. Remarkably, a planetary

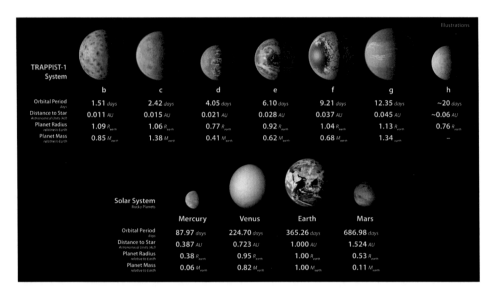

Fig. 9.11   The exoplanet system Trappist-1. (NASA/Caltech)

system with 4 Earth-like planets has been found at Trappist-1. The star is about 40 light years away from us and is located in the constellation Aquarius. The planet Trappist-1 b is too close to the star, but in the spectrum we find an atmosphere containing hydrogen. The planets Trappist-1 e, Trappist-1 f and Trappist-1 g are in the habitable zone. In Fig. 9.11 the most important known sizes of this planetary system are given and below a comparison with planets of our solar system. Some planets of Trappist-1 may also have evolved life on the side facing away from the Sun or the side facing the Sun.

The brightness curves of the individual planets in the system are shown in Fig. 9.12. These data were obtained with the Spitzer telescope. This telescope operates primarily in the infrared, and it was launched into space in 2003 on an orbit that provided protection from solar radiation. In order to observe undisturbed in the infrared, the detectors of the telescope were cooled to 2 K, the coolant (liquid helium) lasted until 2009, after which the detectors operated at a temperature of about 30 K. The telescope was launched in 2003. The telescope was then shut down in 2020. The infrared range is particularly interesting for exoplanet research. Here the contrast between star and planet is lower than in the visible range.

9.3.7   Proxima Centauri

Proxima Centauri (proxima means the nearest) is our more immediate neighboring system. The star Proxima Centauri can only be observed in the southern night sky below the

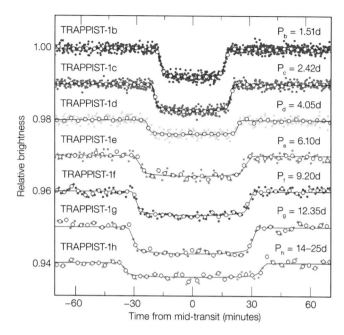

Fig. 9.12 Light curves of the planets of the Trappist-1 system. (NASA/Spitzer)

Fig. 9.13 Comparison of the sizes of the Sun, α and β Centauri and Proxima Centauri. (After D. Benbennick)

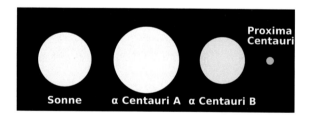

latitude of 27°. It is about 2° away in the sky from the bright stars α and β Centauri and orbits them in about 500,000 years. It is therefore a triple star system (Fig. 9.13). The distance from Proxima Centauri to us is 4.24 light years. The mass is about 0.12 solar masses, the radius 0.15 solar radii, the spectral class M5. In August 2016, a planet Proxima Centauri b was found. It has a little more than one Earth mass and orbits the star in about 11 days. From this planet, α and β Centauri would be seen as bright stars, much brighter than Venus in the sky. Interestingly, from the two bright stars, Proxima Centauri could just be seen with the naked eye because of its faintness. Life seems very unlikely on the planet Proxima Cen b, Proxima Centauri belongs to the group of flare stars where very violent bursts of radiation occur; the short-wave UV and X-rays would wipe out life. Moreover, the planet is bound to the star by tidal action and always shows the same hemisphere to the star.

The Voyager space probe, which travels at a speed of 61,000 km/h, needs 75,000 years for a journey to Proxima Centauri.

9.4   Are We Alone in the Universe?

We have found exoplanets for the first time in the last few decades, so it seems that there will soon be an answer to the burning question of the probability of life elsewhere. Attempts were made even before the discovery of exoplanets to estimate whether we are alone or not.

9.4.1   Drake Equation

Frank Drake used a radio telescope around 1960 to listen to messages from extraterrestrial civilizations. He came up with an equation that makes it possible to estimate how large the number of civilizations in our galaxy or universe could be that are at about the same technological level as we are. It results:

$$N = S f_P n_{pm} f_l f_c L. \tag{9.1}$$

The individual quantities mean:

N	Number of civilization at our technical level or higher.
S	Total number of stars in the milky way. This number is easy to specify: a few hundred billion.
f_P	Number of stars that have a planetary system. This number is still uncertain, but values between 0.5 and 1 are assumed. The value $f_P = 1$ would mean: Every star has a planetary system.
n_{pm}	Number of planets (and moons) that lie in a habitable zone, i.e. where life could exist. In our solar system there are >1 candidates, so we set this number optimistically between 0.1 and 2.
f_l	Number of objects in a system where life has actually evolved. Most biochemists assume that, given the right conditions, life will inevitably evolve. This number is set between 0.01 and 1.
f_c	Number of planets with civilizations at a high technical level. On earth, it took four billion years for such a civilization to develop, almost half the lifetime of our sun. often one puts $f_c = 1$.
L	Indicates the probability that such a civilization still exists today. On earth we have a civilization on a high technical level for about 100 years. If one sets for $L = 10^{-7}$, then this means a life span of a civilization of 1000 years. However, one could also use $L = 10^{-2}$, which means that such a civilization survives 100 million years.

Let's be pessimistic. Using the lowest values for the Drake equation, we are probably the only civilization in the Milky Way. That means communication with other civilizations would be impossible. Still, the universe would be full of life, there are a few hundred billion of galaxies, so a few billion civilizations, even if only one out of 100 galaxies had a planet with advanced civilization.

In the best case scenario, there could be several ten million civilizations in our Milky Way alone.

▸ Even with very pessimistic assumptions, there should be several billion advanced civilizations in the universe.

9.4.2   SETI and Other Projects

SETI means Search for Extraterrestrial Intelligence. You try to listen to the sky for radio signals sent by such civilizations. But at what frequencies should we listen to the sky? One assumes very special frequencies, e.g. the interstellar 21-cm-line of hydrogen. This line should be known to an intelligent civilization. The special thing about SETI, however, is that it is not sponsored by states, but the project lives on private donors as well as a very large number of computer users who download radio data onto their PCs. By means of software running in the background, this data is then automatically examined for special patterns that stand out from the noise. So every computer user on earth can actively contribute to the search. In 1974 a message was sent to the globular cluster M13 with the large Arecibo radio telescope. Due to the distance of M13, however, we have to wait 48,000 years before an answer arrives. The space probes Pioneer 10 and 11, launched in 1972 and 1973, respectively, carry a golden plaque (Fig. 9.14) that schematically shows the solar system with the planets, the hydrogen atom, and a pair of humans. On board the Voyager probes launched in 1977 is a golden record with various voices (Pope, US President, UN Secretary General, etc.). One of the probes will be near a star in 30,000 years, but will still be 1 light year away from it.

Whether these probes will ever be found by an extraterrestrial civilization is considered very unlikely.

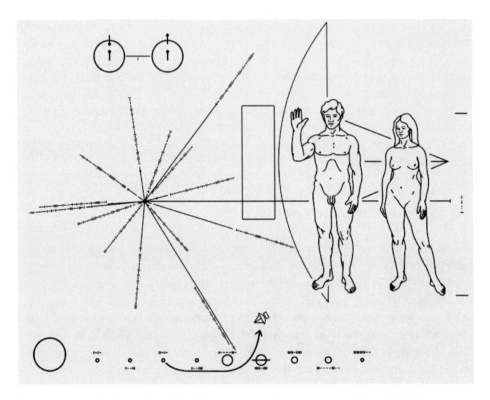

Fig. 9.14 The gold plaque aboard the Pioneer spacecraft, a message to aliens. (NASA)

9.5 The History of the Universe in One Day

In the second chapter of the book, we presented the evolution of the universe when the time scale is projected onto a year. Here we want to show a comparison again: What would the history of the universe look like if the entire evolution from the Big Bang to today took place in just one day? In particular, we will point to the origin of life.

- 0:00: The universe comes into being, big bang.
- 00:00:02, i.e. two seconds after the Big Bang: matter, dark matter, hydrogen and helium already exist; universe is transparent.
- 01:30:00 first quasars galaxies form; probably also our Milky Way; first stars form; massive stars shine only 5–10 s and explode to a supernova; stars like our Sun shine more than 10 h.
- 15:35:00: Our solar system is formed by the collapse of an interstellar gas cloud.
- 15:40:00: Our Earth collides with a Mars-sized planet, the Moon is formed.
- 17:00:00: The first life emerges on Earth; primitive single-celled cyanobacteria.

Fig. 9.15 Inanimate universe compared to the time since life existed on Earth

Life in the universe

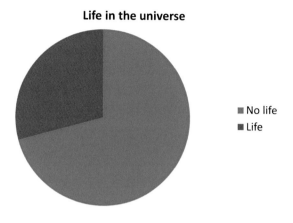

■ No life
■ Life

Fig. 9.16 Comparison of time periods: Earth without life, Earth with primitive life, Earth with humans (this period is so short that it is not noticeable in this plot)

Life on earth

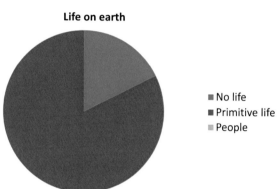

■ No life
■ Primitive life
□ People

- 20:40:00: Multicellular organisms evolve; life becomes more complex.
- 23:00:00: Spread of multicellular organisms, including on land.
- 23:40:00: first dinosaurs.
- 23:52:48: Extinction of the dinosaurs by impact of an asteroid.
- 23:59:35: first ancestors of man.
- 23:59:59.8: modern man.

So the origin of life on Earth didn't happen until 17:00:00 in this model.

Figure 9.15 shows how the periods when there was life on Earth relate to the periods when there was no life. Figure 9.16 shows how the time periods in which there was more highly developed life on Earth relate to the time periods in which there was no life or only primitive life. The length of time that has passed since the first ancestors of man appeared does not matter in comparison with the others.

Index

Printed in the United States
by Baker & Taylor Publisher Services